Managing Products and Services

PS2

HEALTH AND SAFETY

Published for
The National Examining Board for Supervision and Management

by
Pergamon Open Learning

Pergamon Open Learning
An imprint of Butterworth-Heinemann Ltd
Linacre House, Jordan Hill, Oxford OX2 8DP

℞ A member of the Reed Elsevier plc group

OXFORD LONDON BOSTON
MUNICH NEW DELHI SINGAPORE SYDNEY
TOKYO TORONTO WELLINGTON

This unit supersedes the Super Series first edition unit 504 (first edition 1986)

Second edition 1991
Reprinted 1992
Reprinted 1993
Reprinted 1994
Reprinted and updated 1995

A catalogue record for this book is available from the British Library

ISBN book and cassette kit: 0-08-041643-8

The views expressed in this work are those of the authors and do not necessarily reflect those of the National Examining Board for Supervision and Management or of the publisher.

Original text produced in conjunction with the Northern Regional Management Centre under an Open Tech Contract with the Manpower Services Commission.

Design and Production: Pergamon Open Learning

NEBSM Project Manager: Pam Sear
Author: Joe Johnson
First Edition Author: Matthew S. McCall
Editor: Diana Thomas
Series Editor: Diana Thomas

Typeset by BPC Digital Techset Ltd, Exeter
Printed in Great Britain by BPC Wheatons Ltd, Exeter

CONTENTS

USER GUIDE

1   Welcome to the User Guide

Hello and welcome to the NEBSM Super Series second edition (Super Series 2) flexible training programme.

It is quite likely that you are a supervisor, a team leader, an assistant manager, a foreman, a section head, a first-line or a junior manager and have people working under you. The Super Series programme is ideal for all, whatever the job title, who are on or near that first rung of the management ladder. By choosing this programme we believe that you have made exactly the right decision when it comes to meeting your own needs and those of your organization.

The purpose of this guide is to help you gain the maximum benefit both from this particular workbook and audio cassette and also from a full supervisory management training programme.

You should read the whole of this User Guide thoroughly before you start any work on the unit and use the information and advice to help plan your studies.

If you are new to the idea of studying or training by yourself or have never before worked with a tutor or trainer on an individual basis, you should pay particular attention to the section below about Open Learning and tutorial support.

If you are a trainer or tutor evaluating this material for use with prospective students or clients, we think you will also find the information given here useful as it will help you to prepare and conduct individual pre-course counselling and group briefing sessions.

2   Your Open Learning Programme

What do we mean by 'Open Learning'?

Let's start by looking at what is meant by 'Open Learning' and how it could affect the way you approach your studies.

Open Learning is a term used to describe a method of training where you, the learner, make most of the decisions about *how*, *when* and *where* you do your learning. To make this possible you need to have available material, written or prepared in a special way (such as this book and audio cassette) and then have access to Open Learning centres that have been set up and prepared to offer guidance and support as and when required.

Undertaking your self-development training by Open Learning allows you to fit in with priorities at work and at home and to build the right level of confidence and independence needed for success, even though at first it may take you a little while to establish a proper routine.

The workbook and audio cassette

Though this guide is mainly aimed at you as a first time user, it is possible that you are already familiar with the earlier editions of the Super Series. If that is the case, you should know that there are quite a few differences in the workbook and audio cassette, some of which were very successfully trialled in the last 12 units of the first edition. Apart from the more noticeable features such as changes in page layouts and more extensive use of colour and graphics, you will find activities, questions and assignments that are more closely related to work and more thought-provoking.

Guide

The amount of material on the cassette is, on average, twice the length of older editions and is considerably more integrated with the workbook. In fact, there are so many extras now that are included as standard that the average study time per unit has been increased by almost a third. You will find a useful summary of all workbook and cassette features in the charts below and on page vii.

Whether you are a first time user or not, the first step towards being a successful Open Learner is to be familiar and comfortable with the learning material. It is well worth spending a little of your initial study time scanning the workbook to see how it is structured, what the various sections and features are called and what they are designed to do.

This will save you a lot of time and frustration when you start studying as you will then be able to concentrate on the actual subject matter itself without the need to refer back to what you are supposed to be doing with each part.

At the outset you are assumed to have no prior knowledge or experience of the subject and can expect to be taken logically, step by step from start to finish of the learning programme. To help you take on new ideas and information, and to help you remember and apply them, you will come across many different and challenging self check tasks, activities, quizzes and questions which you should approach seriously and enthusiastically. These features are designed not only to make your learning easier and more interesting but to help you to apply what you are studying to your own work situation in a practical and down-to-earth way.

To help to scan the workbook and cassette properly, and to understand what you find, here is a summary of the main features:

The workbook

If you want:	Refer to:
To see which other Super Series 2 units can also help you with this topic	The Study links
An overview of every part of the workbook and how the book and audio cassette link together	The Unit map
A list of the main knowledge and skill outcomes you will gain from the unit	The Unit objectives
To check on your understanding of the subject and your progress as you work thorough each section	The Activities and Self checks
To test how much you have understood and learned of the whole unit when your studies are complete	The Quick quiz and Action checks
An assessment by a third party for work done and time spent on this unit for purposes of recognition, award or certification	The Unit assessment The Work-based assignment
To put some of the things learned from the unit into practice in your own work situation	The Action plan (where present)

The audio cassette

If you want:	Refer to:
To start your study of the unit	The Introduction: Side one
To check your knowledge of the complete unit	The Quick quiz: Side one
To check your ability to apply what you have learned to 'real life' by listening to some situations and deciding what you should do or say	The Action checks: Side two

Managing your learning programme

When you feel you know your way around the material, and in particular appreciate the progress checking and assessment features, the next stage is to put together your own personal study plan and decide how best to study.

These two things are just as important as checking out the material; they are also useful time savers and give you the satisfaction of feeling organized and knowing exactly where you are going and what you are trying to achieve.

You have already chosen your subject (this unit) so you should now decide when you need to finish the unit and how much time you must spend to make sure you reach your target.

To help you to answer these questions, you should know that each workbook and audio cassette will probably take about *eight* to *ten* hours to complete; the variation in time allows for different reading, writing and study speeds and the length and complexity of any one subject.

Don't be concerned if it takes you longer than these average times, especially on your first unit, and always keep in mind that the objective of your training is understanding and applying the learning, not competing in a race.

Experience has shown that each unit is best completed over a two-week period with about *three* to *four* study hours spent on it in each week, and about *one* to *two* hours at each sitting. These times are about right for tackling a new subject and still keeping work and other commitments sensibly in balance.

Using these time guides you should set, and try to keep to, specific times, days, and dates for your study. You should write down what you have decided and keep it visible as a reminder. If you are studying more than one unit, probably as part of a larger training programme, then the compilation of a full, dated plan or schedule becomes even more important and might have to tie in with dates and times set by others, such as a tutor.

The next step is to decide where to study. If you are doing this training in conjunction with your company or organization this might be decided for you as most have quiet areas, training rooms, learning centres, etc., which you will be encouraged to use. If you are working at home, set aside a quiet corner where books and papers can be left and kept together with a comfortable chair and a simple writing surface. You will also need a note pad and access to cassette playing equipment.

When you are finally ready to start studying, presuming that you are feeling confident and organized after your preparations, you should follow the instructions given in the Unit Map and the Unit Objectives pages. These tell you to play the first part of Side one of the audio cassette, a couple of times is a good idea, then follow the cues back to the workbook.

You should then work through each workbook section doing all that is asked of you until you reach the final assessments. Don't forget to keep your eye on the Unit Map as you progress and try to finish each session at a sensible point in the unit, ideally at the end of a complete section or part. You should always start your next session by looking back, for at least ten to fifteen minutes, at the work you did in the previous session.

You are encouraged to retain any reports, work-based assignments or other material produced in conjunction with your work through this unit in case you wish to present these later as evidence for a competency award or accreditation of prior learning.

Help, guidance and tutorial support

The workbook and audio cassette have been designed to be as self-contained as possible, acting as your guide and tutor throughout your studies. However, there are bound to be times when you might not quite understand what the author is saying, or perhaps you don't agree with a certain point. Whatever the reason, we all need help and support from time to time and Open Learners are no exception.

Help during Open Learning study can come in many forms, providing you are prepared to seek it out and use it:

- first of all you could help yourself. Perhaps you are giving up too easily. Go back over it and try again;

- or you could ask your family or friends. Even if they don't understand the subject, the act of discussing it sometimes clarifies things in your own mind;

- then there is your company trainer or superior. If you are training as part of a company scheme, and during work time, then help and support will probably have been arranged for you already. Help and advice under these circumstances are important, especially as they can help you interpret your studies through actual and relevant company examples;

- if you are pursuing this training on your own, you could enlist expert help from a local Open Learning centre or agency. Such organizations exist in considerable numbers throughout the UK, often linked to colleges and other training establishments. The National Examining Board for Supervision and Management (NEBSM or NEBS Management), has several hundred such centres and can provide not only help and support but full assessment and accreditation facilities if you want to pursue a qualification as part of your chosen programme.

The NEBSM Super Series second edition is a selection of workbook and audio cassette packages covering a wide range of supervisory and first line management topics.

Although the individual books and cassettes are completely self-contained and cover single subject areas, each belongs to one of the four modular groups shown. These groups can help you build up your personal development programme as you can easily see which subjects are related. The groups are also important if you undertake any NEBSM national award programme.

Managing Human Resources

HR1	Supervising at Work	HR10	Managing Time
HR2	Supervising with Authority	HR11	Hiring People
HR3	Team Leading	HR12	Interviewing
HR4	Delegation	HR13	Training Plans
HR5	Workteams	HR14	Training Sessions
HR6	Motivating People	HR15	Industrial Relations
HR7	Leading Change	HR16	Employment and the Law
HR8	Personnel in Action	HR17	Equality at Work
HR9	Performance Appraisal	HR18	Work-based Assessment

Managing Information

IN1	Communicating	IN7	Using Statistics
IN2	Speaking Skills	IN8	Presenting Figures
IN3	Orders and Instructions	IN9	Introduction to Information Technology
IN4	Meetings		
IN5	Writing Skills	IN10	Computers and Communication Systems
IN6	Project Preparation		

Managing Financial Resources

FR1	Accounting for Money	FR4	Pay Systems
FR2	Control via Budgets	FR5	Security
FR3	Controlling Costs		

Managing Products and Services

PS1	Controlling Work	PS8	Productivity
PS2	Health and Safety	PS9	Stock Control Systems
PS3	Accident Prevention	PS10	Stores Control
PS4	Ensuring Quality	PS11	Efficiency in the Office
PS5	Quality Techniques	PS12	Marketing
PS6	Taking Decisions	PS13	Caring for the Environment
PS7	Solving Problems	PS14	Caring for the Customer

While the contents have been thoroughly updated, many Super Series 2 titles remain the same as, or very similar to the first edition units. Where, through merger, rewrite or deletion title changes have also been made, this summary should help you. If you are in any doubt please contact Pergamon Open Learning direct.

First Edition	**Second Edition**
Merged titles	
105 Organization Systems and 106 Supervising in the System	HR1 Supervising at Work
100 Needs and Rewards and 101 Enriching Work	HR6 Motivating People
502 Discipline and the Law and 508 Supervising and the Law	HR16 Employment and the Law
204 Easy Statistics and 213 Descriptive Statistics	IN7 Using Statistics
200 Looking at Figures and 202 Using Graphs	IN8 Presenting Figures
210 Computers and 303 Communication Systems	IN10 Computers and Communication Systems
402 Cost Reduction and 405 Cost Centres	FR3 Controlling Costs
203 Method Study and 208 Value Analysis	PS8 Productivity
Major title changes	
209 Quality Circles	PS4 Ensuring Quality
205 Quality Control	PS5 Quality Techniques
Deleted titles	
406 National Economy/410 Single European Market	

The NEBSM Super Series 2 Open Learning material is published by Pergamon Open Learning in conjunction with NEBS Management.

NEBS Management is the largest provider of management education, training courses and qualifications in the United Kingdom, operating through over 700 Centres. Many of these Centres offer Open Learning and can provide help to individual students.

Many thousands of students follow the Open Learning route with great success and gain NEBSM or other qualifications.

NEBSM maintains a twin track approach to Supervisory Management training offering knowledge-based awards at three levels:

● the NEBSM Introductory Award in Supervisory Management;
● the NEBSM Certificate in Supervisory Management;
● the NEBSM Diploma in Supervisory Management;

and competence based awards at two levels:

● the NEBSM NVQ in Supervisory Management at Level 3;
● the NEBSM NVQ in Management at Level 4.

Knowledge-based awards and Super Series 2

The ***Introductory Award*** requires a minimum of 30 hours of study and provides a grounding in the theory and practice of supervisory management. An agreed programme of up to five NEBSM Super Series 2 units plus a one-day workshop satisfactorily completed can lead to this Award. Pre-approved topic combinations exist for general, industrial and commercial options. Completed Super Series 2 units can be allowed as an exemption towards the full NEBSM Certificate.

The ***Certificate in Supervisory Management*** requires study of up to 23 NEBSM Super Series 2 units and participation in group activity or workshops. The assessment system includes work-based assignments, a case study, a project and an oral interview. The certificate is divided into four modules and each one may be completed separately.
A ***Module Award*** can be made on successful completion of each module, and when the assessments are satisfactorily completed the Certificate is awarded. Students will need to register with a NEBSM Centre in order to enter for an award; NEBSM can advise you.

The ***Diploma in Supervisory Management*** consists of the formulation and implementation of a Personal Development Plan plus a generic management core. The programme is assessed by means of a log book, case study/in tray exercises, project or presentation.

The NEBSM Super Series 2 Open Learning material is designed for use at Certificate level but can also be used for the Introductory Award and provide valuable background knowledge for the Diploma.

The **_NEBSM NVQ in Supervisory Management Level 3_** is based upon the seven units of competence produced by the Management Charter Initiative (MCI) in their publication *Supervisory Management Standards* of June 1992. It is recognized by the National Council for Vocational Qualifications (NCVQ) at Level 3 in their framework.

The **_NEBSM NVQ in Management Level 4_** is based upon the nine units of competence produced by MCI in their publication *Occupational Standards for Managers, Management 1 and Assessment Criteria* of April 1991. It is recognized by the National Council for Vocational Qualifications (NCVQ) at Level 4 in their framework.

Super Series 2 units can be used to provide the necessary underpinning knowledge, skills and understanding that are required to prepare yourself for competence-based assessment.

Working through Super Series 2 units cannot, by itself, provide you with everything you need to enter or be entered for competence assessment. This must come from a combination of skill, experience and knowledge gained both on and off the job.

You will also find many of the 47 Super Series 2 units of use in learning programmes for other National Vocational Qualifications (NVQs) which include elements of supervisory management. Please check with the relevant NVQ lead body for information on Units of Competence and underlying knowledge, skills and understanding.

Competence Match
Chart

The Competence Match Chart overleaf illustrates which Super Series 2 titles provide background vital to the current MCI M1S Supervisory Management Standards. You will also find that there is similar matching at MCI M1, Management 1 Standards. This is shown on the chart on page xiii.

For more information about MCI contact:

Management Charter Initiative
Russell Square House
10–12 Russell Square
London
WC1B 5BZ

Progression

Many successful NEBSM students use their qualifications as stepping stones to other awards, both educational and professional.
Recognition is given by a number of bodies for this purpose. Further details about this and other NEBSM matters can be obtained from:

NEBSM Information Officer
The National Examining Board for Supervision and Management
1 Giltspur Street
London
EC1A 9DD

Competence Match Chart MCI M1S

The chart shows matches of Super Series 2 titles with MCI M1S (Supervisory Management) Units of Competence. Titles indicated ● are directly relevant to MCI Units, those marked ◑ provide specific supporting information, and those listed ○ provide useful general background.

NEBSM Super Series 2 Titles   **MCI M1 S Units of Competence (see below)**

		1	2	3	4	5	6	7
PS1	Controlling Output	◑	◑					
PS2	Health and Safety	●	○			○		
PS3	Accident Prevention	●	○			○		
PS4	Ensuring Quality	●	○					
PS5	Quality Techniques	●						
PS6	Taking Decisions	○	○			◑	◑	
PS7	Solving Problems	○	○			◑	●	
PS8	Productivity		◑			●		
PS9	Stock Control Systems		◑					
PS10	Stores Control		◑					
PS11	Efficiency in the Office		◑			◑		
PS12	Marketing	○						
PS13	Caring for the Environment	◑	◑			○	○	○
PS14	Caring for the Customer	◑	○			○		
HR1	Supervising at Work					●	●	
HR2	Supervising with Authority					●	●	
HR3	Team Leading					●	●	
HR4	Delegation				●	●	◑	
HR5	Workteams					●	●	
HR6	Motivating People					●	●	
HR7	Leading Change		◑			●		
HR8	Personnel in Action			●				
HR9	Performance Appraisal				●		●	
HR10	Managing Time		○		○			
HR11	Hiring People			●				
HR12	Interviewing			●	●	◑	●	
HR13	Training Plans				●			
HR14	Training Sessions				●			
HR15	Industrial Relations						●	
HR16	Employment and the Law			○			●	
HR17	Equality at Work			◑			●	
HR18	Work-based Assessment			○	●	●	○	○
FR1	Accounting for Money		●					
FR2	Control via Budgets		●					
FR3	Controlling Costs		●					
FR4	Pay Systems							
FR5	Security	◑	◑					
IN1	Communicating	○	○	○	○	○	○	●
IN2	Speaking Skills	○	○	○	○	○	○	●
IN3	Orders and Instructions	◑				●	●	
IN4	Meetings				○	●	◑	●
IN5	Writing Skills	○	◑		○	◑	○	●
IN6	Project Preparation				○			
IN7	Using Statistics	◑	◑					●
IN8	Presenting Figures	◑	◑					●
IN9	Introduction to Information Technology	◑	◑					●
IN10	Computers and Communication Systems	◑	◑					●

* MCI M1 S Units of Competence

1. Maintain services and operations to meet quality standards
2. Contribute to the planning, monitoring and control of resources
3. Contribute to the provision of personnel
4. Contribute to the training and development of teams, individuals and self to enhance performance
5. Contribute to the planning, organization and evaluation of work
6. Create, maintain and enhance productive working relationships
7. Provide information and advice for action towards meeting organizational objectives

The chart indicates the Super Series 2 titles which provide some useful background information to support MCI M1 (Management level 1) Units of Competence.

Guide

NEBSM Super Series 2 Titles		MCI M1 Units of Competence (see below*)								
		1	2	3	4	5	6	7	8	9
PS1	Controlling Output	△	△							
PS2	Health and Safety	△								
PS3	Accident Prevention	△								
PS4	Ensuring Quality	△	△							
PS5	Quality Techniques	△	△							
PS6	Taking Decisions								△	△
PS7	Solving Problems		△						△	△
PS8	Productivity		△							
PS9	Stock Control Systems	△								
PS10	Stores Control	△								
PS11	Efficiency in the Office	△	△							
PS12	Marketing	△								
PS13	Caring for the Environment	△								
PS14	Caring for the Customer		△							
HR1	Supervising at Work							△		
HR2	Supervising with Authority							△		△
HR3	Team Leading					△	△	△		
HR4	Delegation					△	△	△		
HR5	Workteams					△	△	△		△
HR6	Motivating People							△		
HR7	Leading Change		△							
HR8	Personnel in Action				△					
HR9	Performance Appraisal							△		
HR10	Managing Time									
HR11	Hiring People				△					
HR12	Interviewing				△	△		△		
HR13	Training Plans					△				
HR14	Training Sessions					△				
HR15	Industrial Relations							△		
HR16	Employment and the Law				△			△		
HR17	Equality at Work				△			△		
HR18	Work-based Assessment					△	△			
FR1	Accounting for Money			△						
FR2	Control via Budgets			△						
FR3	Controlling Costs			△						
FR4	Pay Systems									
FR5	Security									
IN1	Communicating							△		△
IN2	Speaking Skills			△				△		△
IN3	Orders and Instructions							△		△
IN4	Meetings							△		△
IN5	Writing Skills			△			△	△		△
IN6	Project Preparation			△			△	△		
IN7	Using Statistics						△	△	△	
IN8	Presenting Figures						△	△		△
IN9	Introduction to Information Technology								△	△
IN10	Computers and Communication Systems								△	△

* MCI M1 Units of Competence

Key Role: Manage Operations — 1. Maintain and improve service and product operations
2. Contribute to the implementation of change in services, products and systems

Key Role: Manage Finance — 3. Recommend, monitor and control the use of resources

Key Role: Manage People — 4. Contribute to the recruitment and selection of personnel
5. Develop teams, individuals and self to enhance performance
6. Plan, allocate and evaluate work carried out by teams, individuals and self
7. Create, maintain and enhance effective working relationships

Key Role: Manage Information — 8. Seek, evaluate and organise information for action
9. Exchange information to solve problems and make decisions

Guide Unit Completion Certificate

Completion of this Certificate by an authorized and qualified person indicates that you have worked through all parts of this unit and completed all assessments. If you are studying this unit as part of a certificated programme, or think you may wish to in future, then completion of this Certificate is particularly important as it may be used for exemptions, credit accumulation or Accreditation of Prior Learning (APL). Full details can be obtained from NEBSM.

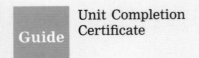

PS2

Health and Safety

. . . .

has satisfactorily completed this unit.

Name of Signatory. . . . .
Position. . . . .
Signature. . . . .

Date . . . .

Official Stamp

Keep in touch

Pergamon Open Learning and NEBS Management are always happy to hear of your experiences of using the Super Series to help improve supervisory and managerial effectiveness. This will assist us with continuous product improvement, and novel approaches and success stories may be included in promotional information to illustrate to others what can be done.

1   NEBSM Super Series 2 study links

Here are the Super Series 2 units which link to *Health and Safety*. You may find this useful when you are putting together your study programme but you should bear in mind that:

- each Super Series 2 unit stands alone and does not depend upon being used in conjunction with any other unit;

- Super Series 2 units can be used in any order which suits your learning needs.

TRAINING PLANS
Training in safe working practices and maintaining a safe environment for themselves and others is an important part of team development. This unit helps identify training needs and plan systematic training.

TEAM LEADING
How to build an efficient and effective workteam whose members are committed to high standards and shared responsibility - essentials of a safe workplace.

ORDERS AND INSTRUCTIONS
Accidents sometimes result from misunderstanding, incomplete instructions and lack of follow up. This unit helps you to ensure your instructions are clear, complete and meet the needs of the situation.

HEALTH AND SAFETY
This unit describes how the law affects health and safety at work and helps you to promote a healthy and safe environment in your work area and among your workteam.

ACCIDENT PREVENTION
This unit examines the law relating to accident prevention, describes the main hazards and causes of accidents in a number of areas and helps you organize and maintain accident precautions.

SECURITY
This unit helps develop practical security policies which are closely aligned to a healthy and safe workplace.

EMPLOYMENT AND THE LAW
This unit covers other aspects of the law related to work.

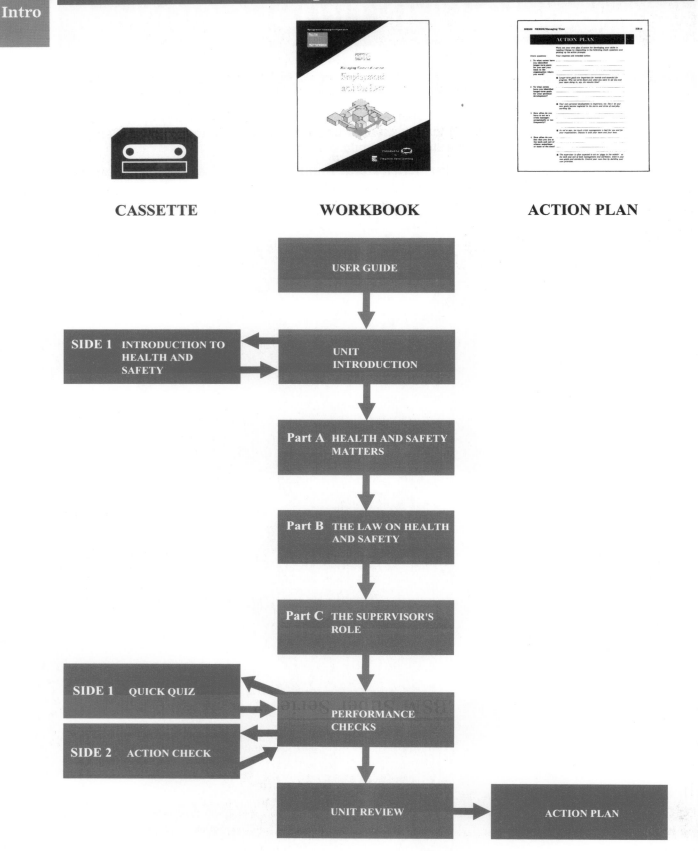

CASSETTE   WORKBOOK   ACTION PLAN

USER GUIDE

SIDE 1 INTRODUCTION TO HEALTH AND SAFETY

UNIT INTRODUCTION

Part A HEALTH AND SAFETY MATTERS

Part B THE LAW ON HEALTH AND SAFETY

Part C THE SUPERVISOR'S ROLE

SIDE 1 QUICK QUIZ

SIDE 2 ACTION CHECK

PERFORMANCE CHECKS

UNIT REVIEW

ACTION PLAN

Supervisors at work have responsibilities for health and safety as individual employees, in taking steps to ensure the safety of themselves and others;

■ as representatives of management, in implementing the employer's policies and procedures; and

■ as team leaders, in helping to ensure that their members take a positive attitude towards health and safety.

In this unit, we will examine a number of aspects of health and safety, particularly from the supervisor's viewpoint.

Before you start work on this unit, listen carefully to Side 1 of the accompanying audio tape which sets the scene for your examination of *Health and Safety*.

In this unit we will:

● consider the importance and the impact of health and safety issues on the working environment;

● look at the responsibilities and duties of people in making and maintaining safe and healthy places of work;

● examine the law related to health and safety;

● investigate the role of the supervisor.

Objectives

When you have worked through this unit you will be ***better able to***:

● recognise the impact of health problems and accidents on working life, and the importance of taking steps to minimise them;

● explain your duties and responsibilities, and those of your team members, in health and safety matters;

● identify the most important legislation related to health and safety and explain the duties imposed by the law on everyone at work;

● play an active part in helping the people in your workplace to remain safe and healthy.

HEALTH AND SAFETY MATTERS

1   Introduction

There are always underlying reasons for accidents and health problems in the workplace, perhaps the most common being a kind of complacency which allows people to believe: 'It can't happen to us!'

Yet it is probably true to say that in almost every workplace there are accidents, or other health-related incidents, simply 'waiting to happen'. These are situations, systems and procedures which are not fully under the control of the people involved, or which depend too much on human ability and judgement to make them safe.

In this part of the unit, we'll set the scene by discussing what health and safety means, the effects and costs of accidents and health problems, and what can be done to minimize their impact.

2   Who's responsible?

- 'Man Killed when Excavation Trench Falls In.'
- 'Asbestos Workers Sue for Compensation.'
- 'Worker Dies after Paint Is Ignited by Gas-Powered Lamp.'
- 'Scaffolding Collapses, Injuring Three.'
- 'Repetitive Strain Injury Affects Factory Workers.'
- 'Investigation into Crash of Chemical Tanker.'
- ''Carelessness at Work Kills,'' Says Safety Expert.'

We've all read headlines like these. It's easy to get the impression that accidents and health problems in places of work occur very frequently. Unfortunately, the statistics are even more frightening.

In regard to accidents alone, it is estimated that

- approximately *ten* people are killed at work *every week* in this country;
- in any year, hundreds of thousands are injured or become ill as a result of accidents;
- for every major injury accident, there are ten incidents involving property damage;
- for every property damage incident, there are twenty 'near misses'.

But it isn't only accidents that can affect the health of workers. There are many other work-related illnesses – for example, resulting from:

- physical and mental stress;
- working in unhealthy conditions;
- handling hazardous substances;
- infections.

The natural reaction of people when they realize these facts is that

'Something should be done about it!'

Activity 1

■ Time guide 3 minutes

Whose job is it to 'do something about' the high rate of accidents at work? Tick which of the following groups you think have a responsibility to do something.

The government ☐

Employers ☐

Employees ☐

Managers ☐

Supervisors ☐

Some other group: _____ ☐

Well, you would be right to tick all the first five boxes. If you added others, such as 'safety officers' or 'The Health and Safety Executive', you were probably right again. You may also have suggested trade unions, and they certainly play a very important role.

You may agree that **everybody** who:

● works, or

● employs people to work, or

● has any control or influence over workplaces,

has a **moral** responsibility to do everything possible to make workplaces safer and healthier.

In fact,

> employers, employees, managers and supervisors all have responsibilities under the health and safety laws.

We'll look at these laws in some detail later in the unit.

Although everybody at work has responsibilities for health and safety, the fact is that ordinary employees look to managers and supervisors for leadership in these matters, as in everything else. And it is only supervisors and managers who have the opportunity to implement safe systems and adequate controls.

The law recognizes these facts. When something goes wrong, it is nearly always managers and supervisors who are deemed to be responsible.

We will return to this theme again later.

3   The costs

The cost of accidents and health problems at work can be measured in financial terms, both to the employer and the injured or sick person.

Extension 1 The Health and Safety Executive (HSE) has published a booklet *The costs of accidents at work* which describes five case studies, analysing the cost of accidents for each. This is well worth reading if you are at all concerned about costs.

Let's look at the employer's position first.

Activity 2

■ Time guide 3 minutes

Spend a few minutes writing a list of the ways in which an accident at work may cause an employer to lose money. Try to think of *three*.

_____

_____

_____

_____

Money may be lost by the employer through:

■ not having the services of the injured person while he or she is unable to work, including perhaps the cost of hiring temporary staff;

■ the disruption to the work of other people;

■ the time spent by the supervisor in training replacement staff and perhaps taking part in an accident enquiry;

■ possible damage to equipment;

■ cost of employer welfare benefits;

■ possible claims for compensation.

Financial losses can be separated into direct and indirect costs, and into insured and uninsured costs. The following HSE diagram lists some of the costs in these categories:

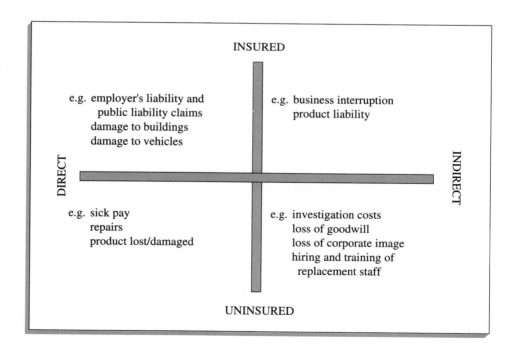

(Based on *The costs of accidents at work*, HSE, Crown copyright 1993.)

7

Of course, the extent of insured losses depends upon the kind of insurance policy held.

The HSE points out that uninsured costs are often far greater than insured costs. In one case, there was a ratio of 36:1 – in other words, the uninsured costs of accidents over the period of monitoring were 36 times as great as the insured costs!

How about the employee's financial position?

Activity 3

■ Time guide 3 minutes

Note how an employee may lose money as a consequence of having had an accident at work.

_____

_____

_____

Here are my suggestions:

■ even with sickness benefit, the person's income may go down, especially if the injury results in a long period of absence from work;

■ in the worst case, he or she may not be able to return to work at all;

■ there may be other costs – travel to and from a medical centre, costs of prescriptions and so on.

Cumulatively, accidents and health problems have an effect on the national economy. The worse the health and safety record of work organizations, the worse off we all are.

The costs in human terms – the physical and mental distress of the people involved and their families – are much more difficult to evaluate, although just as real. If you've known someone who has been badly hurt at work, you'll appreciate how high the human costs can be.

So we have to do all we can to reduce accidents and health-related problems.

4   Accidents and why they happen

Let's start with an important premise:

Accidents are always preventable

Take these three accidents for example:

Jennie went to the tea-room for her mid-day refreshment, where boiling water for making tea was kept in a free-standing electrical boiler. Jennie slipped on the wet floor, instinctively grabbed the boiler for support and pulled it over on top of her. She received extensive burns. The water on the floor had come from the boiler, which had a dripping tap.

Eddie was using a portable electric drill fitted with an abrasive disc to clean some metalwork. As he leant over the work, Eddie's tie became entangled and wrapped itself around the drill chuck. Before he could stop the drill, it had wound itself around the tie, and travelled upwards. The abrasive disc caused Eddie severe facial injuries.

A bath of flammable solvent was being used to clean metal parts. Carl, who was doing this work, was provided with an apron which prevented the solvent getting onto his clothes, but for some reason he chose not to use it. Carl then went to a rest area to light a cigarette and his impregnated clothing caught fire. His life was saved by the quick thinking of a workmate who used a fire extinguisher to put out the fire.

Three different accidents, and apparently three different causes. In the first a dripping tap caused a floor to become slippery. In the second, a man's tie became entangled in a drill. In the third, a man failed to use the protective clothing provided.

These accidents were all preventable. However, it would no doubt be possible to describe fifty or a hundred other work accidents, and each might have a different cause.

So how can we talk about 'accident prevention' in any general way?

Activity 4

■ Time guide 4 minutes

Can you think of one feature that *all* work accidents have in common? If you're stuck for an answer think back to the three incidents above. Ignore the 'technical' causes – which we know were all different – and try to find a more fundamental cause which is common to them all.

_____

_____

As we saw above, the 'technical' causes of different accidents may have nothing in common. So to find a common feature we need to move away from technicalities and think in terms of *people* and *organizations*.

Your response to the question in this Activity may have been:

■ 'poor supervision', or

■ 'inadequate training', or

■ 'lack of instruction', or

■ 'lack of information', or

■ 'inadequate procedures'.

These would all be correct. The majority of accidents result from these failures.

Accidents come as no surprise. Something happens, or is allowed to happen, which ought to have been picked up before. In other words, whatever safety system exists, it is out of control.

So perhaps the best response to this Activity might be:

■ accidents at work are largely caused by systems out of control.

What does this mean?

Safety at work can only be achieved if there are well-organized *systems* of safety. The reasons are that:

● the safety of people at work depends on cooperation between individuals and between groups;

● people at work largely do what they are told to do or what they are allowed to do;

● people at work tend to assume that someone higher in the organization is making sure that the workplace is safe.

A system implies a definite policy or strategy for safety, and clear procedures and rules.

> When accidents occur, it's often because the safety systems are inadequate, or because people have failed to follow the rules.

Later in this section, we will discuss the key elements needed for successful health and safety management.

Let's now look at another couple of accidents and see if we can analyse the causes behind them.

Activity 5

■ Time guide 6 minutes

Jim Dolling operated a powerful machine with moving parts. Jim was very experienced – he had been using this machine for over three years. The machine came fitted with a safety guard, which prevented anything becoming entangled in the gears. The job involved inserting plastic sheets. Sometimes a sheet got jammed. When this happened, the machine had to be switched off and the guard removed so that the sheet could be extricated. After this had happened a few times, Jim never bothered to replace the guard, because it was a time-consuming operation. His supervisor knew about this, and occasionally used to remind Jim that the guard should be in place. Jim told him not to worry, and in fact Jim never had an accident. Then one day Jim went sick, and another man took over the machine. This man caught a loose overall strap in the machine and received serious injuries.

■ What were the 'technical' causes of this accident?

_____

_____

_____

■ Who was responsible? Tick which of the following you think were responsible:

— Jim ☐

— the new operator ☐

— the supervisor ☐

— senior management ☐

— the safety officer – assuming there was one ☐

— the manufacturers of the machine ☐

— the machine maintenance staff ☐

Perhaps you agree with me that:

■ the guard was left off the machine (mainly because it had to be removed frequently and was time-consuming to replace);

■ a less experienced operator used an unguarded machine; and

■ the operator wore loose clothing.

Who was responsible?

■ Jim was: he could – and should – have replaced the guard, especially after reminders by his supervisor.

■ The new operator was: he should have been more careful about his clothing, and could have replaced the guard before using the machine.

■ The supervisor was: he should have gone further than reminding Jim about the guard – he should have insisted it be replaced. Also, he should not have allowed a less experienced operator to work at the machine without a guard.

■ Senior management were, because responsibility passes all the way to the top of an organization.

■ What about the safety officer, assuming there was such a person in this organization? The essential role of a safety officer is to give advice to the organization and its employees: this job could be compared with that of a quality manager. He or she cannot be expected to monitor every faulty condition that might exist.

■ The manufacturers of the machine perhaps were: did they give enough thought to safety when they designed a machine with a guard which needed to be removed frequently and was troublesome to replace?

■ The machine maintenance staff perhaps were: did they maintain the machine well enough, or was it poor maintenance which caused the frequent jams?

It may seem that there are a lot of people involved. You may agree with me that this isn't surprising, because activities at work nearly always involve a number of people. The fact is that the accident might have been avoided *if only* one or more of these people or groups had done a better job.

('If only' is a phrase often heard after an accident ...)

Here's a description of another incident.

Activity 6

■ Time guide 6 minutes

Billy had just started work for his local council on a job training scheme and was to spend his time working for the Parks and Highways Department. On his first day, Billy was given overalls and safety boots by the supervisor, and told to wear them whenever he was at work. Billy disliked the idea of wearing boots, and anyway the boots pinched his toes. The next day he came to work in his favourite training shoes. In the afternoon, Billy had to work with the supervisor to lift some barrels from an open wagon. One of the barrels was so heavy they dropped it. It fell on Billy's foot, breaking all his toes.

This time, imagine you were given the job of investigating this accident. Jot down *two* or *three* important questions you might ask during your investigation.

_____

_____

_____

_____

_____

_____

Some of the questions you might ask are listed here.

■ Was Billy told why it was important to wear the safety boots, and was it emphasized that he wouldn't be allowed to work unless he was wearing them?

■ Were other people doing similar jobs allowed to work in their own footwear? In other words, were the rules normally insisted upon?

■ Did the supervisor notice that Billy wasn't wearing the boots?

■ Was it made clear to Billy that if the boots or overalls didn't fit properly he could ask for a different size?

■ How was it that the barrel was heavier than expected? Should a fork-lift truck or other equipment have been used to unload the wagon?

■ Was Billy (or the supervisor) given training in lifting heavy weights?

You may well have thought of other questions. In an accident investigation like this, the answers frequently lead to the conclusion that more than one factor is involved.

To summarise we can say that:

> * Accidents are always preventable.
>
> * When accidents occur, it's often because the safety systems are inadequate, or because people have failed to follow the rules.
>
> * Accidents nearly always have more than one cause.
>
> * There is often more than one person or group at fault in any accident.

Accident prevention is not the only area of health and safety at work.

Let's look at just a few aspects of health in the workplace which are not related to specific incidents.

5.1
Alcohol, medication and drugs

Activity 7

■ Time guide 3 minutes

What are the rules of your organization regarding alcohol?

_____

_____

Most organizations take a serious attitude towards the consumption of alcohol either during work hours or before coming to work. It is common for this offence to be regarded as gross misconduct, and for employees to be liable to instant dismissal. From the health and safety point of view, it is important to make sure that no one under the influence of alcohol is allowed to

● operate machinery;

● climb ladders or scaffolding;

● drive vehicles;

● do any other work requiring:

    – steadiness of hand;

    – prompt physical reactions;

    – a clear head.

The same rules apply to many forms of medication or drugs, whether taken on prescription or otherwise. Someone taking medication to alleviate the symptoms of hay fever may become drowsy, for instance.

In some workplaces, a supervisor may need to be on the look out for drug and solvent abuse.

5.2
Work-related upper limb disorder

There is an increasing awareness these days of the fact that some jobs can cause temporary or long-term injuries as a result of repeated, forced or awkward movements, or of having to hold part of the body in an unnatural static position.

Many employees in many kinds of job have been affected: typists, VDU operators, stackers and assemblers, butchers, supermarket check-out assistants – the list is very long.

The term 'work-related upper limb disorder (WRULD)' – you may sometimes also hear the term 'repetitive strain injury (RSI)' – includes such medical conditions as

● Tenosynovitis

This is inflammation of the sheath covering tendons, usually the tendons of the wrist or hand. If not treated, tenosynovitis (also known as 'teno') can lead to permanent injury. It is a 'prescribed disease' which means that a person developing it can claim industrial injury benefit and may be able to claim compensation from an employer. As soon as symptoms appear (numbness, tingling or pain during movement), complete rest is recommended.

● Muscle strain

This occurs especially in the muscles of the neck and shoulders. Office workers (typists, VDU operators and so on) are just as likely to suffer from muscle strain as manual workers.

The most common causes of WRULD are:

● incorrect posture;

● too heavy a workload;

● repeated movements, especially if made in a forceful or awkward way;

● having to hold the arms and head in a stiff unnatural posture (while sitting at a keyboard, say) resulting in strain on the shoulder and neck;

● inadequate rest periods.

Activity 8

■ Time guide 3 minutes

From this information and from your own experience and knowledge, can you suggest some actions that a supervisor might take to help prevent such injuries?

_____

_____

_____

_____

I suppose that one positive action is to listen to people who complain about feeling pain when they have been doing a particular job. The worst thing to do is to ignore these symptoms, and to allow the person to continue doing the same job in the same way for a long time.

A supervisor may be able to help by:

■ altering the layout of the work area, so that simpler, more natural movements are possible;

■ providing training;

■ setting up a system of job rotation, so that no one person has to perform the same kind of movements for too long a period;

■ ensuring that furniture is appropriate for the job, that chairs and benches are at a suitable height, for example;

■ ensuring that adequate rest periods are given.

Part A

5.3
Health and display screen equipment

The use of visual display units (VDUs) and other display screen equipment is becoming more and more commonplace these days. Properly used, display screens usually present few health problems. However, because the technology is fairly new, many people now have to sit in front of a VDU all day who have never had to do so previously. It is perhaps not surprising that there are a number of fears and complaints about the effect of VDUs on health.

Let's list some of the possible problems.

Under the Health and Safety (Display Screen Equipment) Regulations 1992, which we will mention again in Part B, employers have an obligation to safeguard the health of employees while operating display screen equipment.

Visual problems

Complaints about 'eyestrain' and other visual problems are perhaps the most common. Two scientific studies have shown that over 60 per cent of people working with VDUs suffer visual problems. (However, the same studies found that almost exactly the same percentage of clerical staff *not* working with VDUs suffered visual problems, also.)

Eye strain and other visual problems should certainly not be ignored. They are made worse by:

● having to focus at the same distance for long periods of time;

● having to read (either screens or documents) which are over- or under-illuminated;

● having to focus at an uncomfortable distance.

Some good advice seems to be that VDU operators should:

● be allowed regular breaks from their work;

● be encouraged to adjust screen brightness controls to a comfortable level;

● be encouraged to adjust the position of the screen to a comfortable distance and height so far as is possible;

● sit so that the light from lamps or windows does not reflect onto the screen.

Problems affecting the tendons and muscles

VDU operators may suffer from injuries and strain affecting the wrists, hands, arms, shoulders, neck and back. These include tenosynovitis, muscle and back strain.

To help avoid such health problems, it is advisable to pay proper attention to layout of the work area and to giving operators regular rest breaks.

Radiation and VDUs

Radiation is the word used to describe electrical and magnetic energy which 'travels' through the air in 'waves'. Radiation may include X-rays, infra-red rays, ultraviolet rays, microwaves and so on. Light and sound are also forms of radiation. The way these forms of radiation differ is in the range of frequencies that they transmit (frequency is the number of waves passing a point in one second).

VDUs may emit radiation in a wide spectrum of frequencies, and you may have read about VDU radiation health hazard 'scares'.

According to the Health and Safety Executive: 'There is no evidence at present that users of VDUs need to take special precautions to protect against radiation emissions.'

**5.4
Fitness for work**

It is obviously important that working people are physically and mentally suited to their jobs, and that they are fit to do them.

Supervisors may sometimes have a difficult task in deciding whether a particular team member is fit to work or fit to do a particular job. Someone returning to work after a period of illness, for example, may need time and help to readjust before being able to perform to full capacity.

When trying to decide whether someone is fit for work, it's probably a good idea to get help from others – the personnel department or company doctor, for example.

Before we go on to talk about the management of health and safety, try the following questions.

Self check 1

■ Time guide 10 minutes

1. Which of the following statements are TRUE, and which FALSE?

   (a) Everyone at work has a responsibility to try to reduce accidents and health problems. TRUE/FALSE

   (b) Accidents happen. They always have and always will. There's nothing much you and I can do about it. TRUE/FALSE

   (c) Costs of accidents may be measured in financial terms, but are immeasurably high in human terms. TRUE/FALSE

   (d) It's important to keep health and safety at work in perspective: not that many people are killed or injured at work. TRUE/FALSE

2. Why do accidents happen? Tick the *two* statements below which describe the *most common* reasons behind accidents at work.

Accidents occur:

   (a) because supervisors don't do enough to ensure the safety of their team members; ☐

   (b) when people 'cut corners' and don't follow laid-down procedures; ☐

   (c) when safety procedures and systems have bee designed or implemented inadequately; ☐

   (d) when guards are left off machines. ☐

continued

3. Say whether you agree, or do not agree, with each of the following statements, and give a brief reason to justify your views.

(a) If an employee is taking drugs, he or she should not be working.

Agree/diasagree because _____

_____

_____

_____

_____

(b) When a team member complains that he or she is suffering pain while working, the supervisor should take some action.

Agree/diasagree because _____

_____

_____

_____

_____

(c) Most people who complain of visual problems when working at VDUs are probably imagining it.

Agree/diasagree because _____

_____

_____

_____

_____

4. Complete the following sentences, using suitable words chosen from the list below.

SYSTEMS   PERSON   UNINSURED   INDIRECT   CAUSE
DIRECT   INSURED   GROUP   PREVENTABLE   CONTROL

(a) Financial losses can be separated into _____ and _____ costs, and into _____ and _____ costs.

(b) Accidents are always _____ .

(c) Accidents at work are largely caused by _____ out of .

(d) Accidents nearly always have more than one .

(e) There is often more than one _____ or _____ at fault in any accident.

Part A

Response check 1

1. (a) Everyone at work has a responsibility to try to reduce accidents and health problems. This is TRUE. Everyone has not only a responsibility, but a duty to try to reduce accidents and health problems.

   (b) Accidents happen. They always have and always will. There's nothing much you and I can do about it. This is FALSE. The first two sentences may be true – even if not very helpful. The third sentence is definitely false.

   (c) Costs of accidents may be measured in financial terms, but are immeasurably high in human terms. This is TRUE.

   (d) It's important to keep health and safety at work in perspective: not that many people are killed or injured at work. This is FALSE. If you believe that 'not that many' are killed or injured, you haven't read the statistics!

2. Looking at each of the statements:

   (a) Accidents occur because supervisors don't do enough to ensure the safety of their team members. This may be true on occasions, but supervisors certainly can't be held responsible for the majority of accidents.

   (b) Accidents occur when people 'cut corners' and don't follow laid-down procedures. This *is* a very common cause of accidents.

   (c) Accidents occur when safety procedures and systems have been designed or implemented inadequately. This is also a very common cause.

   (d) Accidents occur when guards are left off machines. This is one 'technical' cause of accidents involving machines, but can't be said to be a common cause of all types of accident.

3. (a) If an employee is taking drugs, he or she should not be working.

   Certainly some drugs, including those that can cause drowsiness, may not be appropriate in a work situation. However, it would be going too far to ban all drugs.

   (b) When a team member complains that he or she is suffering pain while working, the supervisor should take some action.

   Yes, even if the only action taken is to listen and take note. Supervisors have a duty to ensure that the health of team members does not suffer unnecessarily because of the work they are asked to do.

   (c) Most people who complain of visual problems when working at VDUs are probably imagining it.

   Sometimes you may get that impression. But it might be foolish to assume this is true in every case. Staring at a screen all day is not easy on the eyes, and the problems complained of may be very real.

4. (a) Financial losses can be separated into DIRECT and INDIRECT costs, and into INSURED and UNINSURED costs.

   (b) Accidents are always PREVENTABLE.

   (c) Accidents at work are largely caused by SYSTEMS out of CONTROL.

   (d) Accidents nearly always have more than one CAUSE.

   (e) There is often more than one PERSON or GROUP at fault in any accident.

6   Managing health and safety

Extension 1 In its booklet *Successful Health and Safety Management*, the HSE sets out the key elements to success in managing and supervising health and safety at work. We can do no better than to follow the advice given there.

These key elements are summed up in the diagram below:

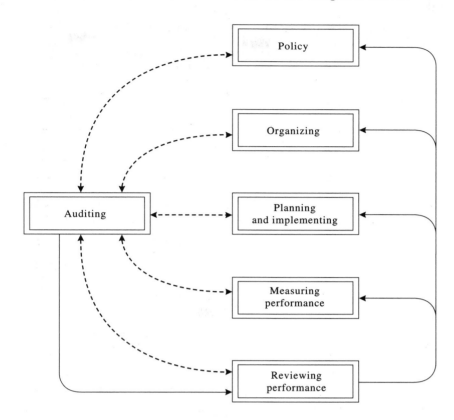

(Diagram based on *Successful Health and Safety Management*, Health and Safety Executive, Crown copyright 1991.)

Let's look briefly at each of these key elements.

6.1 Policy

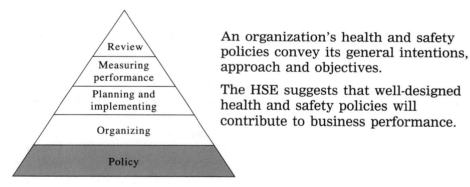

An organization's health and safety policies convey its general intentions, approach and objectives.

The HSE suggests that well-designed health and safety policies will contribute to business performance.

Activity 9

■ Time guide 3 minutes

Can you suggest one way in which a company's health and safety policies could contribute to its business performance?

_____

_____

Policies largely the determine the 'environment' or 'climate' of a work organization. Maybe you agree that people working in 'a healthy environment' are likely to have fewer accidents and less illness, and therefore be off work less often. The result is less disruption, fewer orders lost and a workforce which is happier and more positive. Accidents often cause damage to property, too, which also costs the organization money.

As the HSE says:

> 'The best health and safety policies are concerned not only with preventing injury and ill health . . . but also with positive health promotion which gives practical expression to the belief that people are a key resource.'

> 'The ultimate goal is an organization in which accidents and ill health are eliminated, and in which work forms part of a satisfying life, contributing to physical and mental well-being, to the benefit of both the individual and the organization.'

As a supervisor, you may not feel that you have much influence over your employer's policies, although you will certainly be able to help create the right environment at a local level.

**6.2
Organizing**

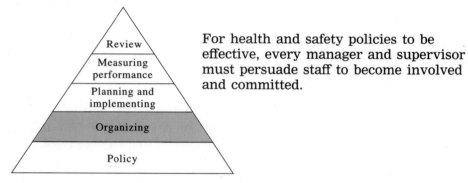

For health and safety policies to be effective, every manager and supervisor must persuade staff to become involved and committed.

Organizing is concerned with the four Cs:

● methods of **control** within the organization;

● the means of securing **cooperation** between individuals, safety representatives and groups;

● the methods of **communication** throughout the organization;

● the **competence** of individuals.

I'll return to the subject of the four Cs later in the workbook.

**6.3
Planning and
implementing**

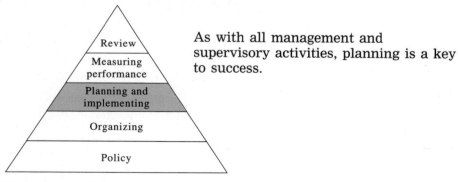

As with all management and supervisory activities, planning is a key to success.

Planning for health and safety involves:

● setting objectives – deciding what you want to achieve;

● identifying hazards;

● assessing risks;

● putting standards into place;

● developing the right culture for safe and healthy working.

You will need to (at least):

● agree with your manager the health and safety targets you and your team are aiming for;

● make sure you understand the law related to the work you do, and comply with it;

● identify the hazards in your workplace, and assess the risks;

● follow agreed safety procedures and rules in planning work tasks;

● plan for dealing with emergencies;

● cooperate with other supervisors, managers and teams.

Activity 10

■ Time guide 10 minutes

Answer the following questions about your own job and place of work, and make a note of any actions you think you should take at this time – such as talking to your manager or your team about the matter.

■ Are you and your team fully aware of the organization's health and safety plans?

| Yes | | No | |

Action: _____

■ Do you think about health and safety before you begin each task?

| Yes | | No | |

Action: _____

■ Are you confident that all the hazards have been identified and all the risks assessed?

| Yes | | No | |

Action: _____

■ Do you have a plan to deal with fires and other possible serious or imminent dangers?

| Yes | | No | |

Action: _____

■ Are your team fully informed about risks and how to control them?

| Yes | | No | |

Action: _____

These are all crucial questions, and you may want to make a note of them, so that you can give more time and thought to them. Your answers will reflect:

● how well you feel you know your organization's health and safety policies and plans;

● your confidence that they are adequate;

● the extent to which you feel the policies and plans have been implemented in your local area.

Part A

6.4
Measuring performance

As a supervisor, you may well be familiar with measuring your team's work performance.

Activity 11

■ Time guide 3 minutes

What are the basic steps involved in measuring performance – that is, finding out how well you or your team have performed?

_____

_____

_____

I hope you agree that measuring performance involves knowing what you want to achieve, and then determining to what extent you have achieved it. For example, if you want to measure production output, you need to know the production level aimed for and then compare the actual figures with this.

The same procedure applies to health and safety performance measurement. You need to set standards and objectives, and then to monitor what is happening – or what has happened.

Extension 1 The Health and Safety Executive, in its leaflet *Five steps to successful health and safety management*, separates monitoring systems into two key components:

● ***active monitoring*** – regular inspections and checks, to find out whether health and safety standards are being followed, and whether they are effective;

● ***reactive monitoring*** – investigating injuries, cases of illness, property damage and near misses, so as to find out how health and safety systems can be improved.

6.5
Reviewing performance and auditing

By monitoring health and safety, you will find out the information you need to review performance, and decide how to improve it.

If your organization carries out audits of health and safety performance and activities, you may well be provided with further information about just how safe your work area is.

Together with your colleagues, you should be able to make use of all this information to improve the reliability and effectiveness of your systems.

Self check 2

■ Time guide 10 minutes

1. Complete the diagram showing the key elements of successful health and safety management:

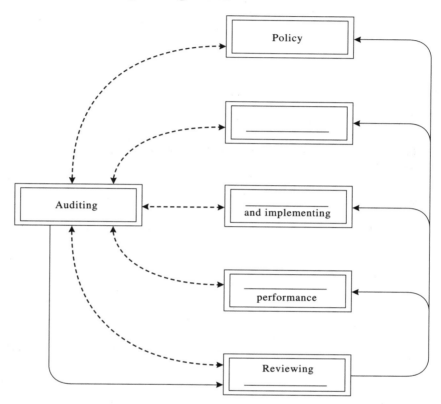

2. Explain in your own words what is meant by the right 'climate' or 'environment' for successful health and safety management in a work organization, and why it is so important. You do not need to write more than two or three sentences.

_____

_____

_____

_____

continued

3. Which of the following statements are TRUE and which FALSE?

   (a) For health and safety policies to be effective, every manager and supervisor must persuade staff to become involved and committed. TRUE/FALSE

   (b) Unlike other types of management and supervisory activity, planning for health and safety is a key to success. TRUE/FALSE

   (c) Two activities involved in the process of health and safety planning are assessing risks and setting objectives. TRUE/FALSE

   (d) Active monitoring means carrying out regular inspections and checks. Reactive monitoring means putting systems into place. TRUE/FALSE

Response check 2

1. The completed diagram is as follows:

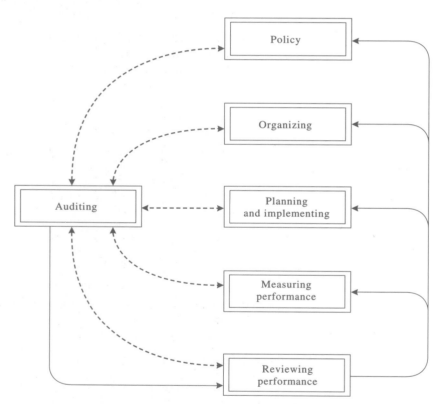

2. The right 'climate' or 'environment' for successful health and safety management is one in which employees at all levels are encouraged to be committed and involved, and in which there is good communication and support. This climate stems from good health and safety policies, and is made effective by good organizing. It is important because it will increase business effectiveness and may save lives.

3. (a) For health and safety policies to be effective, every manager and supervisor must persuade staff to become involved and committed. This is TRUE.

(b) Unlike other types of management and supervisory activity, planning for health and safety is a key to success. This is FALSE: the first word should be 'like' not 'unlike'.

(c) Two activities involved in the process of health and safety planning are assessing risks and setting objectives. This is TRUE.

(d) Active monitoring means carrying out regular inspections and checks. Reactive monitoring means putting systems into place. This is FALSE: reactive monitoring means investigating injuries, cases of illness, property damage and near misses, so as to find out how health and safety systems can be improved.

7   Summary

- The statistics show that the levels of accidents and health problems at work are extremely high.

- Everybody at work has a responsibility to take steps to make the workplace healthier and safer.

- The costs of accidents and health problems may be measured in financial terms, but are immeasurably high in human terms.

- Financial losses can be separated into direct and indirect costs, and into insured and uninsured costs. Often uninsured costs are far greater than insured ones.

- Accidents are always preventable.

- When accidents occur, it's often because the safety systems are inadequate, or because people have failed to follow the rules.

- Accidents nearly always have more than one cause.

- There is often more than one person or group at fault in any accident.

- Possible work-related health problems include:

  – the effects of alcohol, drugs or medication;

  – work-related upper limb disorder;

  – adverse effects from operating display screen equipment.

- Successful health and safety management involves the key elements:

  – policy;

  – organizing;

  – planning and implementing;

  – measuring performance

  – auditing and reviewing.

THE LAW ON HEALTH AND SAFETY

1   Introduction

The law of the land affects all of us, in many aspects of our lives. This is right and proper, because without laws and law enforcement anarchy would reign.

There have been laws on health and safety since early in the nineteenth century. As we will see in this section, the passing of the Health and Safety at Work etc. Act 1974 (the HSW Act) was a significant step forward, and has set up a framework for more recent specific legislation.

We will take a look at the:

- HSW Act;

- new 'six pack' regulations;

- Control of Substances Hazardous to Health Regulations 1988.

2   Why knowing the law is important

This section is devoted to the law on health and safety. You might perhaps be asking yourself whether it is really necessary to give so much time and space to the law. After all, supervisors are not likely to be involved in law suits, are they?

Activity 12

■ Time guide 3 minutes

Try to write down at least one good reason why supervisors ought to understand something about the law on health and safety.

_____

_____

_____

You may have listed one or more of the following reasons:

■ The law affects everyone at work, so supervisors, like everybody else, ought to be aware of laws affecting the individual.

■ Supervisors have special responsibilities, because they have to take steps to protect the health and safety of their team members, as well as their own.

■ Laws of all kinds, and health and safety laws in particular, place demands upon people to act responsibly towards others. It is important to understand what the law expects and demands, for ignorance is no defence when laws are contravened.

There is always the possibility that action could be taken against you, personally, as well as your organization.

■ Health and safety laws provide guidance on minimum standards.

■ Laws are designed to protect people as well as impose demands upon people. Understanding the law can help a supervisor to know the rights of people at work.

Perhaps the best reason is that:

■ knowing the law is part of your job.

3 An overview of health and safety law

There have been many Acts of Parliament over the years related to health and safety at work. Until recently, many old laws were still in force.

When the Health and Safety at Work etc. Act 1974 (the HSW Act) was introduced it was welcomed as a major step forward. The HSW Act applies to all work premises, and sets out clear principles and guidelines for the promotion of health and safety, for both employers and employees.

The HSW Act is described as an 'enabling' Act, because it gives the government powers to introduce regulations and codes of practice on specific health and safety matters. The HSW Act is thus a framework upon which subsequent legislation is based; I will start by reviewing its main points.

Partly as a result of cooperation between the countries in the European Union, some important new Regulations have been introduced. We shall take a look at these. Finally, we will review the Control of Substances Hazardous to Health Regulations 1988 – known as COSHH.

4 The Health and Safety at Work etc. Act 1974

Two of the most significant clauses in the HSW Act read as follows:

Section 2(1): 'It shall be the duty of every employer to ensure, as far as reasonably practicable, the health, safety and welfare at work of all his employees.'

Extract from Section 7: 'It shall be the duty of every employee while at work ... to take reasonable care for the health and safety of himself and of other persons who may be affected by his acts or omissions at work.'

Activity 13

■ Time guide 2 minutes

■ Who in law has duties under the HSW Act, according to these extracts?

_____

_____

I hope you agree that:

■ both employers and employees have duties under the HSW Act.

Let's look at the duties of the employer first.

5	The employer's overall duties under the HSW Act

As we have seen, under the HSW Act, an employer has a duty

> 'to ensure, as far as reasonably practicable, the health, safety and welfare at work of all his employees.'

Activity 14

■ Time guide 2 minutes

■ What 'yardstick' does the law use to determine whether an employer has carried out his or her duties in regard to health and safety at work? (Hint: look for key words in the above extract.)

_____

_____

The key words in the extract from Section 2(1) are '*as far as reasonably practicable*'. This is the 'yardstick' by which an employer's actions will be judged.

To illustrate what is 'reasonably practicable' so far as health and safety is concerned, read the following case.

Two men employed as labourers on a building site are given the job of carrying some panes of glass from one part of the site to another. After a couple of journeys, both the men complain of cuts to their hands, and request that protective gloves be provided. The foreman tells them not to be so 'soft'. The men carry on for a while and then refuse to carry any more of the glass panes.

Activity 15

■ Time guide 3 minutes

In your opinion:

■ is the labourers' refusal to carry on working justified? YES/NO

■ is the foreman acting in a 'reasonably practicable' manner by insisting they carry on working without hand protection? YES/NO

It seems sensible that the men should be provided with hand protection, because they were sustaining actual injuries in doing the work. The foreman is not 'ensuring as far as reasonably practicable' the safety of the men in his employ. In fact, the foreman should have anticipated the need for gloves and ensured they were provided and used.

This case was perhaps not too difficult to make a judgement about. Other situations may not be so straightforward. Expressions such as 'so far as is reasonably practicable' used in the HSW Act are not defined, and only acquire clear meanings through many interpretations by the courts.

According to the Health and Safety Executive:

* To carry out a duty *so far as is reasonably practicable* means that the degree of risk in a particular activity or environment can be balanced against the time, trouble, cost and physical difficulty of taking measures to avoid the risk.

* If these are so disproportionate to the risk that it would be unreasonable for the persons concerned to have to incur them to prevent it, they are not abliged to do so.

* The greater the risk the more likely it is that it is reasonable to go to very substantial expense, trouble and invention to reduce it. But if the consequences and the extent of a risk are small, insistence on greater expense would not be considered reasonable.

* It is important to remember that the judgement is an objective one and size or financial position of the employer are immaterial.

Extension 1 The above extract was taken from '*Successful Health and Safety Management*, HSE Books, October 1993.

**5.1
Duties to
non-employees**

The HSW Act also places an obligation on employers to take care of the health and safety of non-employees.

Activity 16

■ Time guide 3 minutes

Can you suggest *two* groups of people, other than employees, that an employer may have duties towards under health and safety laws?

_____

_____

_____

You may have mentioned:

■ self-employed people or contractors' employees working on site;

■ customers who visit (for instance) shop or garage premises;

■ visiting suppliers;

■ other visitors;

■ the general public living and working outside of the work site.

5.5
Employer's statement of safety policy

We looked at 'policy' as the first step of sound health and safety management in Part A.

The HSW Act states that it is the duty of every employer of five or more people to prepare and keep up to date:

> 'a written statement of his general policy with respect to the health and safety at work of his employees, and the organization and arrangements for ... carrying out that policy, and to bring the statement and any revision of it to the notice of all of his employees.'

This statement of general policy can be considered in three parts. Often the three parts are clearly separated:

● General policy statement

A statement of the organization's commitment to health and safety, and its obligations towards its employees. This statement should also make clear what the duties of employees are.

● Statement of organization

A statement of specific responsibilities: who is responsible for what aspects of health and safety, and how the organization is structured.

● Statement of arrangements

A statement of the specific arrangements for dealing with health and safety matters.

There should be a clear indication of the manner in which these statements are brought to the notice of employees.

Activity 17

■ Time guide 3 minutes

What do you think is gained by making an employer provide a safety policy statement, as described above?

_____

_____

_____

Perhaps you agree with me that one big advantage is that:

■ every employer is compelled to think very carefully about his responsibilities, and what he must do to meet those responsibilities.

Indeed, the employer must:

■ organize and implement safe systems.

As the statement must include details on specific responsibilities and on arrangements for implementing the safety policy, another advantage is that:

■ employees should have a clearer idea of how to act in a safe manner and what to do if something goes wrong.

31

Part B

This brings us to the subject of the responsibilities of employees – we'll look at those next.

6	The employee's duties

Under the HSW Act, employees have a duty:

● to take reasonable care to avoid injury to themselves or to others by their work activities, and

● to cooperate with employers and others in meeting the requirements of the law, and

● not to interfere with or misuse anything provided to protect their health, safety and welfare.

**6.1
Employees' duties
towards themselves**

Activity 18

■ Time guide 4 minutes

Kenny works for a demolition contractor. Kenny's job involves him in having to work high up on the outside of buildings. The nature of the job means that objects fall from above, and that the conditions underfoot can be treacherous.

We know that Kenny's employer has a duty to take all practicable steps to ensure his safety. But what about Kenny himself? What steps should he take to ensure his own safety?

_____

_____

_____

We can perhaps agree that Kenny should do everything possible for his own safety, including:

■ wear all the protective clothing (safety helmet, boots, safety harness, etc.) that is provided for him;

■ check for himself that this equipment is in good condition;

■ make sure he knows how to use all the equipment and machinery he has to deal with;

■ communicate with his workmates so that he is aware of what hazardous conditions exist at the worksite at all times;

■ behave in a sensible and careful manner, and not take unnecessary risks. All employees have a duty to work diligently, i.e. not to omit to do things that should be done to ensure safety.

32

Activity 19

■ Time guide 4 minutes

And what of the safety of others?

Jot down some of the things you would expect a member of your team to do (and perhaps some of the things you would expect the member **not** to do) in order to help to ensure the safety of others.

_____

_____

_____

_____

Your response would have no doubt been related to the kind of job you do. In general, you might expect a team member to do the following.

■ To think of the safety and health of others when carrying out his or her job.

Kenny on the demolition site would be expected to warn his workmates of any hazard he is creating, such as demolishing a wall. In another kind of job, a typist in an office would be expected to make sure that cables, boxes and other obstacles are not a hazard to people walking by.

■ To behave sensibly and responsibly in matters of health and safety.

For instance, it would be irresponsible for someone to cover up a safety notice, or to use a fire bucket for another purpose, or to prop open a fire door which should be kept closed.

■ Not to indulge in 'horseplay' or practical jokes.

The supervisor sometimes has to take care that a 'harmless bit of fun' is not allowed to turn into something more dangerous. A good leader will make plain what is allowed and what isn't.

■ To obey the rules of the organization.

People tend to break safety rules for three main reasons:

– they aren't aware of the rules;

– they don't see any point in the rules;

– the rules impose conflicting restrictions, such as slowing down a process which the person wants to complete as quickly as possible: there is often a great temptation to 'cut corners'.

■ Time guide 3 minutes

Can you think of an instance where someone in your team has been tempted to cut corners in a job, and thereby has compromised on safety?

_____

_____

_____

Depending on the kind of work you are in, you may have suggested some of the following.

■ Not bothering to put on protective clothing:

'I know I should have worn a safety helmet, but I was only going to be out in the yard for two minutes. How was I to know that it would be slippery and that I would fall and crack my head open?' (Man speaking from hospital bed.)

■ Not using the right equipment:

'The step ladder was in use at the time, and I only wanted one item from the top shelf to finish the whole job. Now it looks like I'll be off work for three months.' (Woman on crutches.)

■ Not isolating equipment before working on it:

'Yes, I admit that I should have checked that the electrical power was off before I asked young Peter to open the fuse-box. I was thinking about how much time the interruption was costing us. Now I'll have to live with this for the rest of my life.' (Supervisor at inquiry into fatal accident.)

■ Working on, knowing the risks, and choosing to ignore them for one reason or another:

'The only way to get to the lift control box is to stick your head into the shaft. I suppose we should have shut down the system – but we'd been told that two people were trapped in the lift between floors. We've never had an accident till now. It was a succession of events that caused it. First of all the lift wasn't faulty at all – it was just that one of the doors wasn't shut properly. The trapped people got out, but no one told us. Then someone must have knocked down the warning notice on the ground floor, and somebody else used the lift just at the time Jim was leaning into the shaft. He didn't stand a chance when that balancing weight came down.' (Maintenance engineer talking after accident.)

To sum up:

● Employees have responsibilities under the HSW Act:

– to take care for their own health and safety, and that of their colleagues;

– to cooperate in meeting the requirements of the law, not to interfere with or misuse anything provided to protect their health, safety and welfare.

● People who cut corners endanger themselves and others.

Having looked at the employer's and employee's duties under the HSW Act, let's turn now to some important new Regulations. These supplement the existing law, and in some cases replace older laws.

34

As a result of Directives agreed by the European Union, six new regulations have been approved by Parliament. They are:

● Management of Health and Safety at Work Regulations 1992;

● Workplace (Health, Safety and Welfare) Regulations 1992;

● Manual Handling Operations Regulations 1992;

● Personal Protective Equipment at Work Regulations 1992;

● Provision and Use of Work Equipment Regulations 1992;

● Health and Safety (Display Screen Equipment) Regulations 1992.

I'll give a brief summary of each set of regulations. You don't have to remember all this information, but you may want to take a note of those aspects which you feel particularly relevant to your job and position, and to decide whether to follow up the references, so that you can find out more.

You may also want to bear in mind the fact that, as a supervisor, you represent your employer.

7.1
Management of Health and Safety at Work Regulations 1992 (MHSW)

These regulations help to spell out more explicitly than the HSW Act what the duties and responsibilities of employers are. They apply to almost all kinds of work. According to the HSE:

'Their main provisions are designed to encourage a more systematic and better organized approach to dealing with health and safety.'

Extension 1 The booklet *New health and safety at work regulations* is a practical guide for managers and supervisors under both the HSW Act and the Management of Health and Safety at Work Regulations (MHSW). We have already discussed the recommended management approach contained in the booklet in the first part of our unit.

Specifically, the MHSW Regulations require employers to:

● *assess the risks of the job:*

They must assess the risk to the health and safety of their employees and anyone else who may be affected by their activity, so that the necessary preventive and protective measures can be identified. Employers with more than five employees have to record the significant findings of the assessment.

● *implement necessary measures:*

Any required health and safety measures that follow from the risk assessment must then be put into practice. These would cover planning, organization, control, monitoring and review – in other words, the management of health and safety which you read about in Part A.

● *provide health surveillance:*

Employers have to provide appropriate health surveillance for employees where the risk assessment shows it to be necessary.

● *appoint competent people:*

These are people who will help the employer devise and apply the measures needed to comply with health and safety laws.

● *provide information:*

Employees, together with temporary employees and others in the employer's undertaking, must be given information they can understand about health and safety matters.

● *provide training:*

Employees have to be given adequate health and safety training, and the employer must ensure they are capable enough at their jobs to avoid risks.

Activity 21

■ Time guide 3 minutes

The law imposes a duty on employers to provide any necessary training on safe practices. Make a note of at least *three* aspects of safe practice in which you think the law would expect training and information to be provided.

For example: 'how to work safely in a particular job'.

_____

_____

_____

An employee needs to know, through clear instructions and/or training, everything that concerns personal safety, including:

■ how to work safely in his or her job;

■ what to do if something goes wrong;

■ where to find safety equipment, and how to use it;

■ all relevant legal requirements;

■ what steps he or she needs to take to safeguard the safety of others;

■ any special hazards.

The MHSW Regulations also require employers to:

● *set up emergency procedures;*

● *cooperate with any other employers who share a work site.*

The Regulations further:

● place duties on employees to follow health and safety instructions and report danger;

● extend the current law which requires employers to consult employees' safety representatives and provide facilities for them. Consultation must now take place on such matters as the:

   – introduction of measures that may substantially affect health and safety;

   – the arrangements for appointing competent persons;

   – health and safety information required by law;

   – health and safety aspects of new technology being introduced to the workplace.

Activity 22

■ Time guide 5 minutes

Glance back through the list of requirements of the MHSW Regulations. Which of them do you think you might be particularly involved in, in your job as supervisor?

_____

_____

_____

_____

Supervisors may be expected to participate in any or all aspects covered by the MHSW Regulations. In particular, you may have noted:

■ implementing specific measures, following risk assessment;

■ providing information and training for your team;

■ helping to set up emergency procedures.

Don't forget that, to your team, you **are** the employer.

**7.2
Workplace (Health,
Safety and Welfare)
Regulations 1992**

These Regulations replace a total of 38 pieces of older law, including parts of the Factories Act 1961 and the Offices, Shops and Railway Premises Act 1963. They cover many aspects of health, safety and welfare in the workplace and apply to all places of work *except*: means of transport, construction sites, mines and mineral exploration sites, fishing boats, and work on agricultural or forestry land away from main buildings.

Activity 23

■ Time guide 4 minutes

As you read through the headings below, tick the boxes of any subjects which are relevant to your own job and you would like to find out more about.

Make a note here of the action you will take to do this.

I will find out more about the items I have ticked by:

_____

_____

They stipulate general requirements for working conditions, related to:

■ *the working environment* – including:

☐ temperature;

☐ ventilation;

☐ lighting;

☐ room dimensions;

☐ suitability of workstations and seating;

☐ outdoor workstations, such as weather protection.

continued

■ *safety* – including:

☐ safe passage of pedestrians and vehicles;

☐ windows and skylights (safe opening, closing and cleaning);

☐ glazed doors and partitions (use of safe material and marking);

☐ doors, gates and escalators (safety devices);

☐ floors (their construction, obstructions and slipping and tripping hazards);

☐ falls from heights and into dangerous substances, and falling objects.

■ *welfare facilities* – including:

☐ toilets;

☐ washing, eating and changing facilities;

☐ clothing storage;

☐ seating;

☐ rest areas (and arrangements in them for non-smokers);

☐ rest facilities for pregnant women and nursing mothers.

■ *housekeeping* – including:

☐ maintenance of workplace, equipment and facilities;

☐ cleanliness;

☐ removal of waste materials.

Existing workplaces have until 1st January 1996 to comply with these regulations.

Extension 1 The Workplace (Health, Safety and Welfare) Regulations 1992 are listed in the HSC booklet: *Workplace health, safety and welfare – Approved Code of Practice.* The booklet explains the meaning of the regulations, and its associated code of practice, giving specific practical guidance on its implementation.

**7.3
Manual Handling
Operations Regulations
1992**

These regulations cover the lifting and manoeuvring of loads of all types. They require the employer to:

● consider whether a load must be moved, and if so, whether it could be moved by non-manual methods;

● assess the risk in manual operations and (unless it is very simple) make a written record of this assessment;

● reduce the risk of injury as far as is reasonable practicable.

Extension 1 The Manual Handling Operations Regulations 1992 are explained in the HSE booklet *Manual handling – Guidance on Regulations.*

**7.4
Personal Protective
Equipment at Work
(PPE) Regulations 1992**

Personal protective equipment (PPE) includes eye, foot and head protection equipment, safety harnesses, life jackets and so on. Employees have to:

● ensure this equipment is suitable and appropriate;

● maintain, clean and replace it;

● provide storage for it when not in use;

● ensure that it is properly used;

● give employees training, information and instruction in its use.

Extension 1 Further information about the PPE Regulations is given in the HSE booklet *Personal protective equipment at work – Guidance on Regulations*.

**7.5
Health and Safety
(Display Screen
Equipment)
Regulations 1992**

In the first part of the unit, we discussed the problems of VDUs and other display screen equipment, and listed some appropriate steps that can be taken. These Regulations put into law the employer's duties regarding the operation of such equipment by employees. Employers have to:

● assess and reduce the risks from display screen equipment;

● make sure that workstations satisfy minimum requirements;

● plan to allow breaks or change of activity;

● provide information and training for users;

● give users eye and eyesight tests and (if need be) special glasses.

Extension 1 If you would like to learn more about the display screen equipment regulations, the HSE booklet *Display screen equipment work – Guidance on Regulations* is the best source of information. An extract from this document is shown overleaf.

SUBJECTS DEALT WITH IN THE SCHEDULE

① ADEQUATE LIGHTING

② ADEQUATE CONTRAST, NO GLARE OR DISTRACTING REFLECTIONS

③ DISTRACTING NOISE MINIMISED

④ LEG ROOM AND CLEARANCES TO ALLOW POSTURAL CHANGES

⑤ WINDOW COVERING

⑥ SOFTWARE: APPROPRIATE TO TASK, ADAPTED TO USER, PROVIDES FEEDBACK ON SYSTEM STATUS, NO UNDISCLOSED MONITORING

⑦ SCREEN: STABLE IMAGE, ADJUSTABLE, READABLE, GLARE/ REFLECTION FREE

⑧ KEYBOARD: USABLE, ADJUSTABLE, DETACHABLE, LEGIBLE

⑨ WORK SURFACE: ALLOW FLEXIBLE ARRANGEMENTS, SPACIOUS, GLARE FREE

⑩ WORK CHAIR: ADJUSTABLE

⑪ FOOTREST

**7.6
Provision and Use
of Work Equipment
Regulations 1992**

The definition of 'work equipment' is very wide, and includes a butcher's knife, a combine harvester, and even a complete power station. Employers must:

● take into account working conditions and hazards when selecting equipment;

● ensure equipment is suitable for use, and is properly maintained;

● provide adequate instruction, information and training.

Organizations have until 1997 to upgrade older equipment.

Extension 1 The Provision and Use of Work Equipment Regulations 1992 are explained in the HSE booklet *Work equipment – Guidance on Regulations.*

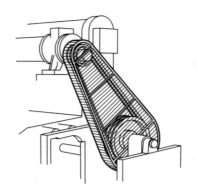

Fixed enclosing guard – an illustration from *Work equipment –
Guidance on Regulations*.

8   The COSHH Regulations

One other set of regulations which has been introduced fairly
recently, and which is extremely important, is the ***Control of
Substances Hazardous to Health Regulations*** 1988 – known as
COSHH.

The COSHH Regulations introduce a new legal framework for
controlling hazardous substances at work.

They cover virtually all substances which can affect health (***except***
asbestos, lead, materials producing ionising radiation and substances
below ground in mines, which all have their own legislation).

Activity 24

■ Time guide 3 minutes

Can you think of any substances which could possibly affect the health of people where you
work?

_____

_____

_____

Unfortunately there are many substances which are hazardous.
Wherever you work – in a factory, in a workshop, in a quarry, on a
farm or garden, in a laboratory, a storage area, or even in an office –
hazardous substances might exist. They may include:

■ anything brought into a workplace to be worked on, used or
   stored: these may include corrosives, acids or solvents;

■ dust and fumes given off by a work process;

■ finished products or residues from a work process.

Anything very toxic, toxic, corrosive, harmful or irritant comes
under COSHH. Examples are chemicals, agricultural pesticides, dusts
and substances containing harmful micro-organisms.

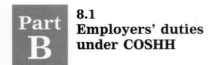

8.1
Employers' duties under COSHH

Under the COSHH Regulations, employers have to:

- **determine the hazard** of a substance which is used by the organization;

- **assess the risk** to people's health from the way the substance is used;

- **prevent anyone being exposed** to the substance, if possible;

- if exposure cannot be prevented, decide how to **control the exposure** so as to **reduce the risk**, and then establish effective controls;

- **train and inform** employees, so that they are aware of the hazards and how to work safely;

- if necessary, **monitor the exposure** of employees and provide health surveillance.

8.2
Supervisors and COSHH

Activity 25

> ■ Time guide 5 minutes
>
> From what you've read so far, can you suggest how supervisors can play a part in helping an employer comply with the COSHH Regulations, and so reduce the risk to employees from hazardous substances? Try to list *two* or *three* positive actions which might be taken by supervisors.
>
> _____
>
> _____
>
> _____

- Probably the most important role supervisors can play is in informing and training the workteam about the hazards and the correct procedures to be followed. We'll look at the kind of information and training needed in a moment.

- Supervisors can and should also ensure that safety procedures are followed, and set a personal example in following the correct procedures consistently and carefully.

- Supervisors are in a good position to assess the likely behaviour of people when they have to deal with hazardous situations, and so can assist an employer in determining what people do and what they might do.

- If protective clothing and/or emergency facilities are provided, it is often the supervisor's job to ensure that these items are available when they're needed, and are properly maintained.

- Supervisors should ensure that only substances are used whose risks have been assessed, and that team members have been trained in the safe handling of these.

Let's now look at the aspects of training and providing information to staff.

A supervisor will need to make sure of the following.

● Team members should understand the hazards.

All suppliers are compelled by law to label hazardous substances. Everyone using such a substance needs to be trained to read and understand container labels and to follow the supplier's advice. More detailed advice is usually available on suppliers' 'data sheets', which list all the hazards of a substance, the effects of exposure and methods of treatment.

● Team members should understand how risks are controlled.

Procedures for controlling risks must be clearly laid down.

● Team members should understand the precautions they have to take.

Precautions and procedures will almost certainly need to be demonstrated.

● Team members should understand what to do in case of emergency.

Emergency procedures need to be demonstrated and practised.

Activity 26

■ Time guide 3 minutes

Looking back over that list of supervisors' responsibilities, make a note of at least three aspects you would want to include in training and development of your staff. For example, you might decide that 'My team members need to be shown the procedures for . . . '

_____

_____

_____

_____

_____

_____

Extension 2 In the Extension you will find a list of publications which give further information about the COSHH Regulations.

9   Other laws

Although we haven't room to cover them in this unit, there are other laws which you should be aware of. Two of them are:

● Electricity at Work Regulations 1989;

● Noise at Work Regulations 1989.

In addition, there may be industry-specific laws which you should be aware of. Your employer may well be able to provide you with information.

Extension 3 The address and telephone number of the Health and Safety Executive is given in this Extension.

It is the job of the Health and Safety Inspectorate to enforce the
health and safety laws.

> The Health and Safety Inspectorate have wide-reaching
> powers. They include:
>
> * the right to enter and inspect any premises, at any
>   time, where it is considered that there may be dangers
>   to health or safety;
>
> * the right to be accompanied by any duly authorized
>   person, such as a policeman or a doctor;
>
> * the rights to aquire into the circumstances of accidents;
>
> * the right to require that facilities and assitance be provided
>   by anyone able to give them;
>
> * the right to make statements;
>
> * the right to require that areas be left undisturbed;
>
> * the right to collect evidence, take photographs, make
>   measurements, etc.;
>
> * the right to take possession of articles;
>
> * the right to require the production of books and
>   documents.

To enforce certain actions, inspectors can:

* issue a *prohibition notice*, which stops – with immediate effect –
  people from carrying on activities which are considered to involve
  a risk of serious personal injury;

* issue an *improvement notice*, which compels an employer to put
  right conditions which contravene the law within a specified time
  period;

* initiate *prosecutions*, especially in the case of repeated, deliberate
  or severe offences.

It should go without saying that supervisors and other managers are
expected to give their full cooperation to the enforcing authorities.
The liability for personal prosecutions is very real.

An employer can appeal against an improvement notice or a
prohibition notice. Here is an example of the case one company put
up against a prohibition notice. The prohibition notice was issued to
prevent a cutting machine being used because a safety guard had
been removed.

*The guard was removed to enable the machine to cope with an
oversize order which was successfully completed. When the guard
was removed the electronic cut-out mechanism which would normally
stop the machine running without the guard was damaged.*

*The manufacturers of the electronic components for this type of
guard have gone out of business, and it will take some time to find a
suitable alternative, although the company is making every effort to
do so.*

*To have a cut-out mechanism made specially would be very
expensive.*

*The company appeals against the prohibition notice on the grounds
of cost and difficulty.*

Activity 27

■ Time guide 4 minutes

Imagine you have to make a judgement on this appeal. You understand that cost and difficulty is an important consideration for any organization, but your main concern is that of safety.

Would you agree that the prohibition should be lifted, given the circumstances? YES/NO

Give a brief reason for your answer.

_____

_____

_____

I would say that, in spite of the cost and difficulties, there is not sufficient reason to lift the prohibition notice. Safety must come first. If the company were to use the machine without a guard, or with a guard that could be removed easily because there is no cut-out mechanism, someone might be seriously injured.

It is in fact very difficult to make a successful appeal against a prohibition notice or an improvement notice.

11   Safety representatives and committees

Everyone has a part to play in health and safety matters. It seems sensible for an employer, therefore, to encourage employee participation in this area.

In this section, we'll take a brief look at the role of safety representatives and safety committees in health and safety.

The regulations covering safety representatives and safety committees are included in Section 15 of the HSW Act.

Extension 1 If you are interested in this subject, you may want to read the Health and Safety Commission booklet *Safety representatives and safety committees.*

**11.1
The safety
representative**

A safety representative is someone appointed by a recognized trade union to represent employees on health and safety matters at work. Because he or she needs to be familiar with the hazards of the workplace and the work being done, safety representatives are usually people with two or more years' experience.

Safety representatives have three main functions. The first is:

● to take all reasonably practicable steps to keep themselves informed.

Activity 28

■ Time guide 4 minutes

What kind of information do you think an employees' representative on health and safety would need in order to do a good job?

_____

_____

_____

Safety representatives would surely need to be familiar with:

■ what the law says about the health and safety of people at work, and particularly the people they represent;

■ the particular hazards of the workplace;

■ the measures which are needed to eliminate these hazards, or to cut down the risk from them;

■ the employer's health and safety policy, and the organization and arrangements for putting that policy into practice.

The other main functions and rights of safety representatives are to:

● encourage cooperation between their employer and employees so that:

– measures can be developed and promoted to ensure the health and safety of employees;

– the effectiveness of these measures can be checked;

● bring to the attention of the employer any unsafe or unhealthy conditions or working practices, or unsatisfactory welfare arrangements.

Activity 29

■ Time guide 4 minutes

Knowing the functions of a safety representative, you may be able to work out the kind of activities involved. Jot down *two* possible activities, if you can.

_____

_____

_____

As you may have mentioned, safety representatives will usually be involved in:

■ talking to employees about particular health and safety problems;

■ carrying out inspections of the workplace to see whether there any real or potential hazards which haven't been adequately addressed;

■ reporting to employers about these problems and other matters connected to health and safety in that workplace;

■ taking part in accident investigations.

Inspections and reports should be formally recorded in writing.

11.2
Safety committees

An employer is legally obliged to set up a safety committee after receiving written requests to do so from two safety representatives.

Typically, a safety committee would be composed of

- a full-time or part-time safety officer – if there is one;
- a works engineer;
- safety representatives;
- a company doctor – if there is one;
- a senior executive of the organization.

The safety committee would:

- review the organization's health and safety rules and procedures;
- study statistics and trends of accidents and health problems;
- consider reports and information received from health and safety inspectors;
- keep a watch on the effectiveness of the safety content of employee training.

Let's look at the kind of accident statistics that would be collected in an organization.

Activity 30

■ Time guide 5 minutes

Read through the following set of accident statistics and try to spot three facts that might be of interest to a safety committee.

For instance, one such fact would be that almost all accidents were minor in nature.

_____

_____

_____

| Period | | Date of accident | Date of engagement | Name, Clock No. and Department | Sex | Age | Occupation | Description of Accident | Nature if injury | Absent days | Code |
From 23.03.94	To 19.04.94										
		25.03.94	30.09.88	J.P. Peters Clock No. 85023 Blownware factory	F	41	Inspector	While inspecting glass in the Blownware factory a flask exploded causing injury.	Cut right forearm.	6	1/14
		09.04.94	10.01.94	C. K. Rush Clock No. 91025 Packing Dept.	F	22	Packer	While packing glass and using stapler machine she felt pain in her neck.	Pain in neck.	5	14
		19.04.94	26.02.89	J.C. Isoz Clock No. 86012 Deptford Cec. Cen.	M	32	Packer	While packing glass he developed pains in both arms and back.	Pain in arm.	10	14
		26.03.94	13.10.90	S. J. Ruffle Clock No. 87222 Deptford Dec. Cen.	F	42	Packer	While lifting glass from crate she strained her back.	Back strain.	15	5
		23.03.94	05.03.84	I. T. Hones Clock No. 91126 Packing Dept.	M	35	Packer	While packing glass, a roll of shrink wrap material, standing on end, fell over and hit foot.	Bruised right toe.	13	5
		10.04.94	27.02.84	A. L. Carvell Clock No. 84076 Deepdale Dept.	F	37	Inspector	While opening cartons to inspect contents, she cut finger causing injury.	Very small cut to bend of small finger of left hand.	2	1/14
		03.04.94	18.03.94	J. Y. Blincowe Clock No. 84321 Plant and Services	M	62	Steel erector	While walking outside of steel erector shop, he twisted his ankle.	Injury to right outer ankle.	9	14
		23.03.94	20.02.94	D. M. Hussein Clock No. 89101 Receiving Stores	M	25	Labourer	While loading laundary onto trailer, he caught his knee on edge of trailer.	Pain right knee	16	4
		07.04.94	23.08.82	J. Austin Clock No. 7934 Transport	M	48	Driver	While trailer was being parked at Deptford parking area, he was jammed between unit and wall.	Fractures to right collar bone, right arm and right foot.	27	5
		13.04.94	19.09.76	G. I. K. Shemwell Clock No. 73211 Transport	M	39	Fork-lift driver	While loading pallets into trailer in Pressware factory, he felt pain in his back.	Pain in lumbar region of back.	5	14

A safety committee may have noted that:

- most accidents were minor in nature;

- the inspecting and packing of glass seems to have resulted in several injuries (though only one cut);

- two of the injuries happened during loading operations;

- people seem to spend a lot of time off work for apparently minor injuries;

- three new members of staff were involved in accidents.

This short section should have given you an idea of the rights and functions of safety representatives and safety committees.

You may also want to note the following point of law. Under the Trade Union Reform and Employment Rights Act 1993, all employees, regardless of their length of service, have a right to complain to an industrial tribunal if they are dismissed or otherwise victimized for:

- carrying out any health and safety activities for which they have been designated by their employer;

- performing any functions as an official or employer-acknowledged health and safety representative or safety committee member;

- bringing a reasonable health and safety concern to their employer's attention in the absence of a representative or committee who could do so on their behalf;

- leaving their work area or taking other appropriate action in the face of serious and imminent danger.

In the next Part of the Unit we will look at reporting accidents, and see how accident statistics are compiled.

Self check 3

- Time guide 10 minutes

1. Fill in the blanks in the following sentences with suitable words:

   (a) The law expects employers to ensure the health, _____ and welfare at work of all its employees, as far as is _____ practicable.

   (b) The law expects employees to take _____ care to avoid injury to themselves or to _____ by their work activities.

   (c) Knowing the _____ is part of a supervisor's job.

   (d) The Health and Safety Inspectorate have the _____ to enter and _____ any premises at any time, where it is considered there may be dangers to health or safety.

   continued

2. Which of the following statements is TRUE, and which FALSE?

The HSW Act:

(a) places duties on employees and employers, but not supervisors. TRUE/FALSE

(b) is the only important law on health and safety. TRUE/FALSE.

(c) requires employers to make a judgement about a risk balanced against the time, trouble, cost and physical difficulty of taking measures to avoid the risk. TRUE/FALSE

(d) requires all employers of five or more people to prepare a written statement of general policy on health and safety. TRUE/FALSE

3. Which of the following are requirements on employers imposed by the law?

(a) To provide information and training about health and safety matters. ☐

(b) To cooperate with other employers on the same site. ☐

(c) To assess risks and implement necessary measures which the assessment shows to be necessary. ☐

(d) To monitor the health of employees at risk. ☐

(e) To be responsible for devising laws on health and safety. ☐

(f) To provide working conditions that ensure employees are never uncomfortable. ☐

(g) To provide suitable and adequate toilet facilities. ☐

(h) To ensure that employees never have to handle anything manually. ☐

(i) To ensure that personal protective equipment is suitable and appropriate. ☐

(j) To plan to allow breaks or change of activity for display screen operators. ☐

4. The Management of Health and Safety at Work Regulations (MHSW) mentions several specific steps required of employers. Fill in the blanks with appropriate words:

The MHSW Regulations require employers to:

● _____ the risks of the job;

● _____ necessary measures;

● _____ health surveillance;

● _____ competent people;

● _____ information;

● _____ training;

● _____ emergency procedures;

● _____ with any other employers who share a work site.

continued

5. Fill in the blanks in the following sentences with suitable words taken from the list below:

EXPOSURE TRAIN CONTROL HEALTH HAZARD
SUBSTANCE MONITOR INFORM EXPOSED RISKS

Under the COSHH Regulations, employers have to:

(a) determine the _____ of a _____ which is used by the organization;

(b) assess the _____ to people's _____ from the way the substance is used;

(c) prevent anyone being _____ to the substance, if possible;

(d) if exposure cannot be prevented, decide how to _____ the _____ so as to reduce the risk, and then establish and maintain effective controls;

(e) _____ and _____ employees, so that they are aware of the hazards and how to work safely;

(f) if necessary, _____ the exposure of employees and provide health surveillance.

6. Which *two* of the following do you think are *the most common* reasons for people breaking safety rules at work.

(a) They are just anti-social. ☐

(b) They are unaware of the rules. ☐

(c) The rules conflict with other priorities, including getting the job done. ☐

(d) They break the rules to show they are brave. ☐

Response check 3

1. (a) The law expects employers to ensure the health, SAFETY and welfare at work of all its employees, as far as is REASONABLY practicable.

(b) The law expects employees to take REASONABLE care to avoid injury to themselves or to OTHERS by their work activities.

(c) Knowing the LAW is part of a supervisor's job.

(d) The Health and Safety Inspectorate have the RIGHT to enter and INSPECT any premises at any time, where it is considered there may be dangers to health or safety.

2. The HSW Act:

(a) places duties on employees and employers, but not supervisors. This is FALSE: the HSW Act imposes duties on all people at work.

(b) is the only important law on health and safety. This is FALSE – as you will know if you have read this Part of the Unit!

(c) requires employers to make a judgement about a risk balanced against the time, trouble, cost and physical difficulty of taking measures to avoid the risk. This is TRUE.

(d) requires all employers of five or more people to prepare a written statement of general policy on health and safety. This is TRUE.

3. All except (e), (f) and (h) are requirements of the law.

4. The MHSW Regulations require employers to:

- assess the risks of the job;
- implement necessary measures;
- provide health surveillance;
- appoint competent people;
- provide information;
- provide training;
- set up emergency procedures;
- cooperate with any other employers who share a work site.

5. Under the COSHH Regulations, employers have to:

(a) determine the HAZARDS of a SUBSTANCE which is used by the organization;

(b) assess the RISKS to people's HEALTH from the way the substance is used;

(c) prevent anyone being EXPOSED to the substance, if possible;

(d) if exposure cannot be prevented, decide how to CONTROL the EXPOSURE so as to reduce the risk, and then establish and maintain effective controls;

(e) TRAIN and INFORM employees, so that they are aware of the hazards and how to work safely;

(f) if necessary, MONITOR the exposure of employees and provide health surveillance.

6. You were asked your opinion, so strictly speaking there are no right and wrong answers. Looking at each of the points in turn:

(a) They are just anti-social. This is NOT a very common reason. For the most part, people at work are conscious of their responsibilities towards others.

(b) They are unaware of the rules. This IS a common reason. It is often the job of the supervisor to make sure people know the rules.

(c) The rules conflict with other priorities, including getting the job done. This IS a common reason, as we discussed.

(d) They break the rules to show they are brave. This is NOT a common reason, although no doubt it does happen.

12   Summary

- The law on health and safety is important to supervisors for a number of reasons, including the facts that they may be liable to prosecution, and that knowing the law is part of their job.

- Under the Health and Safety at Work Act 1974 (the HSW Act) employers have duties to:

  – ensure, as far as reasonably practicable, the health, safety and welfare at work of all their employees;

  – have regard to the safety of non-employees at work;

  – prepare a general policy statement with respect to the health and safety of employees.

- Employees have a duty to:

  – take reasonable care for their own safety and not to hurt others;

  – cooperate with employees and others in meeting the requirements of the law;

  – not to interfere with or misuse anything provided to protect their health, safety and welfare.

- Six new regulations, applicable to most workplaces and covering a wide range of health and safety provisions, are:

  - Management of Health and Safety at Work Regulations 1992;

  - Workplace (Health, Safety and Welfare) Regulations 1992;

  - Manual Handling Operations Regulations 1992;

  - Personal Protective Equipment at Work Regulations 1992;

  - Provision and Use of Work Equipment Regulations 1992;

  - Health and Safety (Display Screen Equipment) Regulations 1992.

- The COSHH Regulations provide a legal framework for controlling hazardous substances at work.

- It is the job of the Health and Safety Inspectorate to enforce the health and safety laws.

- A safety representative is someone appointed by a recognized trade union to represent employees on health and safety matters at work.

THE SUPERVISOR'S ROLE

1   Introduction

We've looked at a number of aspects of health and safety at work. Now we concentrate more on the role of the supervisor.

As usual, the supervisor is at the sharp end of things. He or she is often the one who has to get across the message of health and safety, and the one who has to deal with things first hand when something goes wrong.

In this section we'll look at:

- the duties of supervisors;
- the conditions in the workplace;
- preventing accidents;
- dealing with accidents;
- workplace inspection.

2   How much supervision?

The Health and Safety Executive has specific advice on supervision of health and safety. The booklet *Successful Health and Safety Management* includes the following statement:

'Adequate supervision complements the provision of information, instruction and training in ensuring that the health and safety policy of an organization is effectively implemented and developed.'

But how much supervision is adequate?

Activity 31

■ Time guide 4 minutes

Assume some members of your team are performing a task involving an element of risk. Should you watch over them closely, should you leave them to look after their own safety – or something in between?

Put a tick against any of the following statements which coincide with your views. Add comments, if you wish, to explain your answer.

■ I'd always watch them closely, if possible. ☐

■ I'd always leave them to their own devices. ☐

■ It would depend on how much risk there was involved. ☐

■ It would depend on how experienced and competent the people were. ☐

■ It depends on whether we understood all the risks. ☐

■ It depends on what instructions I got from my superiors. ☐

My comments

_____

_____

The response you gave will depend upon your work situation to some extent. However, I think it would be sensible to make your decision based on the circumstances at the time. Supervisors are expected to learn as much as possible about what the risks are and to take appropriate action. The following HSE diagram is a useful guide:

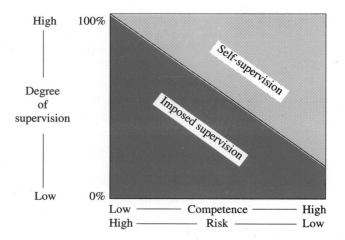

(From *Successful Health and Safety Management*, published by HSE, Crown copyright.)

This diagram shows that the amount of supervision required will depend upon the risk – and what is known about the risk – and the competence of the people involved.

Of course, this means making a judgement. If your team members are young or inexperienced, or you know the risks are high, you will want to make sure they get a lot of support. If the risks are low, and the team has carried out the task many times before, you may feel that they can supervise themselves.

If you aren't sure, then don't take chances. Get advice – talk with your boss or the safety officer.

| 3 | The duties of a supervisor |

Now we'll go on to discuss the duties of supervisors in matters of health and safety.

Activity 32

■ Time guide 5 minutes

Perhaps you'd like to start by listing some of your duties as a supervisor in the area of health and safety. For instance, one duty you will certainly have is to comply with organization rules on health and safety, and try to ensure your team does the same.

Jot down *two* or *three* other general duties you have.

_____

_____

_____

_____

Compare your list with mine.

As a supervisor you will have many health and safety duties, including (usually) the following.

■ To make yourself familiar with your organization's rules on health and safety, and make sure your workteam members are familiar with them and understand them.

■ To comply with those rules yourself and, as far as possible, ensure that workteam members comply with them.

■ To ensure that all equipment and the workplace is maintained so that it is safe to use.

■ To ensure that appropriate and well-maintained protective clothing is available to all members who need it, when they need it, and that it is used.

■ To provide training in safe methods of working, especially to young and inexperienced people.

■ To give adequate instructions about safe methods and procedures.

■ To carry out safety inspections of your work area.

■ To help in accident investigations.

■ To take remedial actions when you see or learn of something that could impair health or safety.

Perhaps you recall the 'four Cs' mentioned in Part A:

I said there that organizing comprises:

■ methods of *control* within the organization;

■ the means of securing *cooperation* between individuals, safety representatives and groups;

■ the methods of *communication* throughout the organization;

■ the *competence* of individuals.

Broadly speaking, the role of a supervisor in the organization's efforts to achieve control, cooperation, communication and competence consists of:

● leading the team;

● managing the task.

As a team leader, part of your job is to encourage cooperation in pursuing health and safety goals. You may need to:

■ encourage participation by all team members, by coaching and counselling;

■ help to make the team aware of the risks involved in the tasks they perform;

■ work together to eliminate risks, or control them more effectively.

As a task manager, you will be expected to ensure specific health and safety objectives are met. Your role may include:

● communicating direction and guidance down and up the management chain;

● providing help and support;

● setting a good example;

● giving training in health and safety skills;

● planning to achieve your objectives;

● monitoring to ensure that standards are being met and maintained.

Obviously, we won't have room in this section to cover all the possible circumstances and conditions which might arise in every kind of job situation.

You may find the following diagram helpful, however. It starts with perhaps the most fundamental question supervisors are faced with:

● 'Am I being reasonable in asking this person to do this job at this time?'

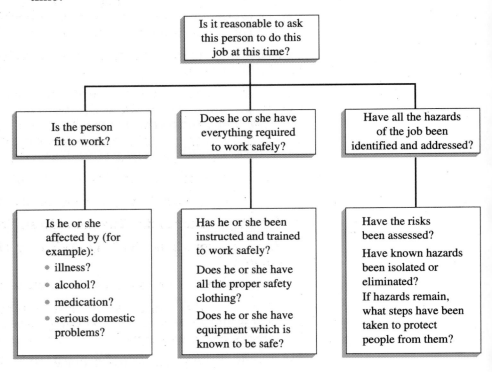

3.1
Sticking by the rules

Sometimes rules exist in a particular workplace which may seem inappropriate or unfair. It isn't unknown for managers and supervisors to 'turn a blind eye' when certain rules are broken.

However, where health and safety are concerned, 'turning a blind eye' to rule-breaking may well have very serious results (perhaps even a blind eye...). The best advice seems to be:

● if you see a safety rule being broken, never turn a blind eye.

What action should you take?

Activity 33

■ Time guide 3 minutes

Imagine you see someone in your team ignoring a safety rule. You point it out to them. Next day you see the same member breaking the same rule. What do you do now?

_____

_____

_____

Perhaps you agree with me that, before taking any disciplinary action, you ought to consider the reason for the member failing to obey the rule. In the last section, we listed three main reasons why people tend to break safety rules:

● they aren't aware of the rules;

● they don't see any point in them;

● the rules impose conflicting restrictions, such as slowing down a process which the person wants to complete as quickly as possible: there is often a great temptation to 'cut corners'.

So the first thing to work out is: 'Why does he (or she) persist in breaking the rules?'

You've already told the person about the rule, so he or she knows that it should be obeyed. Does the member understand the reason for the rule, and what might happen if the rule continues to be broken – perhaps injury or worse? Persuading people of the value of rules can be a difficult job. You may get reactions such as the following.

● 'I'm used to this job. And I've never had an accident.'

● 'It's my neck – what are you worried about?'

● 'I don't need ear-muffs. The noise doesn't bother me.'

● 'Rules are for idiots. I'm fully skilled – that means I know what I'm doing!'

● 'These goggles make me look like something from outer space.'

Another possibility, which we've already discussed, is that there are conflicting requirements, such as being paid a bonus for getting the job done quickly, when the safety rules slow things down.

Having done your best to iron out the reasons for not obeying a safety rule, you may get to the point of insisting that the rule be followed, and instigate disciplinary proceedings if the member persists in disobeying. But, I would suggest, this is a last resort. Enforcing safety rules using some form of punishment is really an admission that the organization has failed to get the message across.

Perhaps the most constructive approach a supervisor can take is to:

● set a good example;

● give every possible encouragement to obey the safety rules;

● make safety a high priority instead of something that lurks somewhere in the background;

● keep monitoring the health and safety performance of your team and workplace.

4   A healthy and safe workplace

Conditions of work can often be the 'hidden' causes of accidents and health problems.

Activity 34

■ Time guide 6 minutes

Make a list of at least *five* possible conditions in a workplace which might lead to accidents or health problems.

One example is poor lighting.

_____

_____

_____

_____

_____

You might have mentioned any of the following:

■ poor lighting, such as lack of light, or glaring or flickering lights;

■ too much or too little heating;

■ inadequate ventilation;

■ an untidy workplace;

■ a dirty workplace;

■ wet or slippery floors;

■ poor drainage;

■ uneven walkways;

■ inadequate space for easy movement;

■ trailing cables or other tripping hazards;

■ lack of handrails or guards where people might fall from floor edges;

■ protruding corners of furniture or tools presenting a hazard to passers-by;

■ emergency exits blocked;

■ icy or frosty conditions making outside walking areas slippery;

■ poorly designed furniture or workspace.

This list is by no means complete. You may recall from Part B that there are regulations – the Workplace (Health, Safety and Welfare) Regulations 1992 – and a Code of Practice which cover working conditions.

Let's just look at one or two of the conditions in more detail.

Activity 35

■ Time guide 3 minutes

Why would an untidy workplace present extra hazards?

_____

_____

_____

You may agree with me that untidiness has both a possible direct and indirect effect.

Untidiness may mean items left lying about which could cause falls or other accidents and injuries. For instance, a broom or a litter-bin in the wrong place can cause someone to trip over; trying to avoid obstacles can cause people to bump into things.

Indirectly, untidiness may result in a general atmosphere of disorderliness, with a consequent lowering of standards of work and behaviour. Even the clutter of paper on a supervisor's desk can give team members the impression that tidiness doesn't matter!

Taking another item on our list, what might be the effects of poor heating?

Activity 36

■ Time guide 3 minutes

How can poor heating result in accidents or health problems?

_____

_____

_____

My comments would be that:

■ uncomfortable heating – whether too hot, too cold or perhaps draughty – can be disturbing to people trying to work, which could cause accidents through loss of concentration. It might also cause them to catch cold;

■ working in very hot or very cold workplaces can be stressful. People under stress are more prone to accidents.

5   Preventing accidents

We've discussed in this unit a number of ways in which accidents may be caused, and conditions which may lead to a higher rate of accidents.

Now let's look at some case incidents of accidents where supervisors were involved.

Lionel Grey had just started a new job in the warehouse of a large supermarket after being unemployed for some time. Lionel's supervisor, Chris Maybank, was busy on the morning he started, and simply gave Lionel the first job that came up, manually unloading crates of tinned goods from a lorry in the loading bay.

Lionel was keen to give a good impression and worked very hard. While doing the job he felt a twinge in his back, but ignored it and carried on working. The next day, the pain was much worse, but he was determined not to let it bother him – he'd been out of work for too long to give up so easily. Chris gave him some more heavy manual work to do and forgot all about him. The next thing Chris knew was that an ambulance had been called for Lionel, who was doubled up with pain. Poor Lionel was diagnosed as having a slipped disc, and had to spend several months off work.

Activity 37

■ Time guide 4 minutes

Do you think the supervisor was at fault in this case? What might Chris have done differently, which could have possibly prevented this accident?

Jot down your views, briefly.

_____

_____

_____

You may have felt that Lionel's problem was nothing to do with Chris the supervisor. After all, he only gave Lionel some routine work – which was what he was paid to do.

But new starters need special treatment. Also, someone who is over-eager to work hard may require as much careful supervision as someone who shows no interest in the job at all. Certainly Lionel should have received some introduction to the work, including training in lifting heavy weights. For instance, there must be a sensible number of cases to carry at one time; someone eager to please like Lionel may well have tried to take too many in one load. And there's a skill in manually moving heavy objects which has to be learned, like any other skill. Had Lionel ever done this kind of work before? Was he fit enough to do this work? Looking back on the incident, Chris may have asked himself these kind of questions. At least he'll know better next time!

Extension 1 The '_Safe moves poster No. 6: Safe way: lifting and carrying_' by the HSE can serve as a useful reminder about the safest ways to handle heavy objects manually.

There was a rule in the office that the new recruits made the tea. Glenda had joined a couple of weeks before and had made tea morning and afternoon for the others. Then one afternoon the kettle wouldn't work any more – the heating element had burned out. Glenda asked what she should do. Her supervisor, Debrah, said: 'Oh, I've an idea. We've still got the old kettle in the basement. I'll fetch it up.' When the old kettle arrived, Debrah said: 'It needs a bit of a clean up, and it leaks a bit. Give it a try, Glenda.' Glenda cleaned up the kettle, filled it up and plugged it in. As soon as she switched on, she got an electric shock which threw her across the room. It turned out that the earth lead was faulty and the live side was touching against the metal body of the kettle.

Activity 38

■ Time guide 4 minutes

Was Debrah at fault here? What could she have done to prevent the accident?

_____

_____

_____

Again, we have a situation where there is a new recruit involved, so that gives us our first clue. New starters always need extra supervision. And I think Debrah should have had more sense than to take an old discarded electrical appliance and ask someone to 'try it out' without having it tested first. Perhaps we should go further back in time and ask why a kettle which was faulty (it was leaking) was kept as a potential hazard for the next unwary user.

Rick Parrish was called out to look at a problem with a job he had given to two of his team. The job was the repair of a tank containing chemicals. They all knew the work was dangerous and Rick had emphasized the precautions they must take. When Geoff, one of the workmen on the job, came into his office to explain the problem, Rick was preparing for a meeting. He said: 'OK, I've just got time to take a quick look.' He then put on his helmet and protective overalls. He looked down at his shoes, thinking 'I ought to change into boots, but I'll only be out there a minute. I might be late for my meeting if I waste any more time.'

As Rick stood over the tank fittings the men were working on, a valve suddenly burst and the chemical came flooding out of the tank. The liquid lapped over the top of Rick's shoes and he was severely burned.

I won't ask you what Rick did wrong in this case – it's too obvious. Even he knew that he ought to have worn the boots that were provided for jobs like these.

This case reminds us that supervisors and managers are among the most vulnerable people at work. They are often so busy thinking of the safety of others that they sometimes forget about their own safety. Yet, as in this case, a supervisor needs first and foremost to set an example for safety – it's far more effective than hectoring others for not being safety-conscious.

A large proportion of accidents happen during maintenance work, because hazards are often much harder to identify.

At the Highbury works, the management were becoming more aware of health and safety, following a couple of serious accidents in the industry. As part of a safety improvement programme, a new instruction was issued, advising everyone working on moving machinery to wear safety goggles.

Tim Bellweather thought the idea was silly. He could see the point if the operation involved the creation of dust or flying particles, but Tim's team was engaged in bending and cutting processes. He'd never known of anyone having an accident involving the eyes.

One morning a box of goggles were left in Tim's office, with a note saying 'Please instruct each member of your team to wear goggles during all machining operations.'

Tim took the goggles into the workshop, dropped the box on a bench and shouted over the noise: 'You've all got to wear these from now on. Don't say I didn't tell you!'

Activity 39

■ Time guide 4 minutes

What was wrong in this situation? Who was to blame? Jot down your thoughts, briefly.

_____

_____

_____

You may agree that Tim was at fault in merely paying lip-service to the instruction about wearing safety goggles. As a supervisor, his job was to set an example and to give instruction in the use of the safety equipment.

However, it seems that Tim wasn't persuaded himself about the value of goggles in the job his team were doing. Did management give Tim any training? Did they tell him about the reasoning behind the move?

Supervisors have an important part to play, but they need support too.

6 Dealing with accidents

What should you do in the event of an accident?

What you should *not* do is to wait until an accident occurs before you think about how you would deal with it.

6.1
Before an accident

Activity 40

> ■ Time guide 5 minutes
>
> What actions could and should be taken in your workplace to **prepare** for possible accidents? Jot down *two* actions that could be taken. (Don't assume that you would have to take these actions on your own, without assistance and support.)
>
> _____
>
> _____
>
> _____
>
> _____
>
> _____
>
> _____

See how far your ideas agree with mine.

■ The first step is to make a *plan* for dealing with each kind of accident that might occur. If you know your job and your workplace well, you're probably in a good position to list the types of accident that could take place.

Some possible types of accident are:

Fire	?		Explosion	?
Flood	?		Poisioning	?
Fall	?		Electrocution	?
Chemical spill	?		Machinery entanglement	?
Lifting accident	?		Radiation leak	?

■ The next step is to *tell people* what to do, such as:

– what might happen;

– where they should go for safety;

– how to raise the alarm;

– how to call the emergency services;

– who the people are who can give first aid, and where they are;

– where the safety equipment is kept;

– how they use the safety equipment;

– how to shut down machinery and electrical equipment;

– how to make an area safe;

– who would normally control an incident.

**6.2
Immediately after an
accident**

Activity 41

■ Time guide 3 minutes

Can you think of some good advice to give someone arriving at the scene of an accident?

_____

_____

_____

Perhaps you suggested:

■ 'Don't panic!'

■ 'Give first aid if possible.'

■ 'Raise the alarm and tell the emergency services.'

■ 'Make sure the area is safe, or keep everyone away.'

These would all be correct. A good five-point plan to follow would seem to be:

> * make the area safe;
> * get first aid to the injured;
> * get help;
> * don't move anything unnecessarily;
> * report the accident.

It's this last point – reporting an accident – which we'll look at next.

**6.3
Reporting an accident**

You will remember when we discussed safety committees that studying accident statistics is an important part of a committee's job.

Of course, these statistics are compiled from accident reports. Often it is the supervisor who fills out the report after talking to those involved.

It isn't only the major catastrophes that need to be reported. Every incident, however minor, should be formally recorded.

Activity 42

■ Time guide 3 minutes

There's often a good deal of reluctance to report accidents. Can you think of *two* reasons why this would be so?

_____

_____

_____

People often consider that:

■ the incident is too trivial to require a report;

■ they might be considered 'soft' by their colleagues if they reported the matter;

■ they would lose working time by reporting it;

■ they have had other accidents, and are afraid of being thought 'accident prone'.

As a supervisor you may have to do your best to encourage members of your team to report all accidents. The reason is that a minor incident may turn out to have more serious repercussions.

Activity 43

■ Time guide 5 minutes

Look at these two descriptions of accidents and write down why you think it was important that the accident should be reported in each case.

■ A member of your workteam slips on the floor in the canteen. She isn't hurt, but tells you that she's heard it said that other people have slipped in the same area.

_____

_____

_____

■ Somebody's eye is hurt by the end of a swinging rope. At the time the eye is just sore, and the injured person 'prefers not to make a fuss'.

_____

_____

_____

■ In the first case, although the woman wasn't hurt, it is evident that there may be a problem with a slippery floor because others are rumoured to have slipped over in the same area. Unless each one of these incidents is reported, it will be difficult to assess the problem. Even if it were the first incident of the kind, there has to have been a reason why it happened. If the hazard isn't identified and eliminated, someone else might fall, with more serious consequences.

■ Any injury, especially one to an eye, **must** be reported. (A blow to the eye is particularly hazardous, and should receive immediate medical treatment, as it is possible for the eye to bleed internally and result in blindness.)

The three main reasons for reporting accidents at work are:

● to help to prevent further accidents. Accident investigations can lead to new safety measures being introduced, so preventing further occurrences;

● to enable compenzation claims to be made. Unless an accident is reported shortly after it happens, it may be difficult to prove at a later stage what took place, or whether the damage or injury actually occurred at the time and location claimed. Often, injuries or other effects don't appear until later;

● so that special precautions can be taken. Accident surveys often show trends, indicating that certain kinds of accident are occurring (say) in a certain way or at a certain time. This can give the organization the opportunity to take appropriate action.

Extension 4 In the Extension, you'll find a sample accident report form.

7   Workplace inspection

A supervisor needs to be sure that the area under his or her control is free from hazards, so far as possible.

One way to do that is to carry out regular and systematic inspections, using formal health and safety checklists.

Two suggested checklists are given below, for fire prevention and office safety. Read these through, and see how far they are relevant to your own needs. There is plenty of space to add check items of your own.

FIRE PREVENTION CHECKLIST			
Date and time of inspection			
ITEM	**Please tick**		**Remarks**
	Yes	No	
Are all fire extinguisers accessible?			
Are all fire alarm points clearly marked and unobstructed?			
Are all fire doors unobstructed?			
Are all sprinkler heads unobstructed?			
Are all fuse boxes in good order and locked?			
Has the fire alarm been tested recently?			
Are all flammable liquid containers marked flammable?			
Are all flammable liquid containers stored correctly when not in use?			
Have all staff been told exactly what to do if they see a fire?			

GENERAL OFFICE SAFETY CHECKLIST			
Date and time of inspection			
ITEM	**Please tick**		**Remarks**
	Yes	No	
Are floors clear of obstructions?			
Is all equipment safely situated?			
Are the floors clean, dry and firm underfoot?			
Are offices clean?			
Are offices generally free from waste paper?			
Are tops of filing cabinets free from waste loose paper?			
Are all entrances and exits clear?			
Is there unobstructed access to all work areas?			
Are offices well ventilated?			
Are cables properly routed?			
Are power points undamaged and not overloaded?			
Is the area free from protuding corners or anything that could present a hazard to a passer-by?			

Activity 44

■ Time guide 10 minutes

Now prepare a checklist for hygiene, using the same format.

To start you off, I've written in the first two items.

CHECKLIST FOR HYGIENE			
Date and time of inspection			
Item	**Please tick**		**Remarks**
Is the area cleaned regularly	Yes	No	
Are washbowls clean?			

One possible checklist is shown on the next page. Don't worry if it doesn't agree with yours. If you start to use checklists like these, you will probably soon begin to notice things that have escaped your attention before; you can then add these to your checklist.

You could devise checklists for 'machinery safety', 'dangerous substances', 'transport and handling', 'maintenance', or any other subject that is useful and appropriate in your own work area.

CHECKLIST FOR HYGIENE			
Date and time of inspection			
Item	**Please tick**		**Remarks**
Is the area cleaned regularly?	**Yes**	**No**	
Are washbowls clean?			
Are overalls and protective clothing regularly cleaned?			
Is there sufficient hot water, soap and barrier cream in the washrooms?			
Are clean towels provided?			
Are toilets regularly cleaned?			
Are rest rooms clean?			
Is the area free from any signs of vermin?			

Self check 4

■ Time guide 10 minutes

1. Which of the following statements are TRUE, and which FALSE?

   (a) Supervisors should never turn a blind eye to safety. TRUE/FALSE

   (b) Safety rules are so important that anyone breaking them should always be punished. TRUE/FALSE

   (c) Safety rules are important, but anyone breaking them should never be punished. TRUE/FALSE

   (d) Safety should be a high priority in any supervisor's job. TRUE/FALSE

2. Which *three* of the following groups would you say were the *most* vulnerable to accidents at work?

   (a) New starters.

   (b) Experienced workers.

   (c) Supervisors and managers.

   (d) Untrained workers.

   (e) Office workers.

continued

3. Which **three** of the following statements are the **best** advice to a supervisor who wants to be prepared for a possible accident?

(a) Ask yourself which kinds of accident might occur, based on your experience and on statistics from your industry. ☐

(b) Put up safety posters to raise safety consciousness. ☐

(c) Send them on courses on safe working practices. ☐

(d) Tell your people what to do in the event of an accident. ☐

(e) Make sure that the proper safety equipment is available. ☐

4. Write down **three** actions you may need to take at the scene of an accident.

_____

_____

_____

Response check 4

1. (a) Supervisors should never turn a blind eye to safety. This is TRUE.

(b) Safety rules are so important that anyone breaking them should always be punished. This is FALSE: it is more important to find out why the rule is being broken than to hand out punishment routinely.

(c) Safety rules are important, but anyone breaking them should never be punished. This is FALSE, also. If someone knowingly breaks the rules, punishment **may** be the only option.

(d) Safety should be a high priority in any supervisor's job. This is TRUE.

2. The groups **most** vulnerable out of those listed are:

(a) new starters;

(c) supervisors and managers;

(d) untrained workers.

3. The best advice seems to be:

(a) Ask yourself which kinds of accident might occur, based on your experience and on statistics from your industry.

(d) Tell your people what to do in the event of an accident.

(e) Make sure that the proper safety equipment is available.

The other two pieces of advice:

(b) 'Put up safety posters to raise safety consciousness' and

(c) 'Send them on courses on safe working practices'

are both useful, but may not actually help very much if an accident should occur.

4. The actions to be taken at the scene of an accident include:

● making the area safe;

● getting first aid to the injured;

● getting help;

● not moving anything unnecessarily.

- The extent of supervision required in a particular situation will depend on an assessment of the risks, taking into account the type of work and the experience of the team members.

- The duties of a supervisor include:

  – *complying* with the health and safety rules of the organization;

  – *helping* team members to comply with the rules.

- It is more important to discover *why* safety rules are being broken than to hand out punishment.

- Constructive advice for a supervisor is to:

  – set a good *example*;

  – give every possible *encouragement* to obey the safety rules;

  – make safety a high *priority* instead of something that lurks somewhere in the background;

  – keep *monitoring* the health and safety performance of your team and workplace.

- *Conditions of work* can often be the 'hidden' causes of accidents and health problems.

- Among the more *vulnerable* groups to accidents at work are *new recruits* and *supervisors*.

- *Before a possible accident*, a *plan* needs to be made for dealing with it.

- *After* an accident, a good five point plan is to:

  – make the area safe;

  – get first aid to the injured;

  – get help;

  – don't move anything unnecessarily;

  – report the accident.

1   Quick quiz

Well done – you have completed the text part of the unit. Now listen to the questions on the second side of the audio tape. If you are not sure about some of the answers, check back in the text before making up your mind.

Write down your answers in the space below.

1 _____

   _____

2 _____

   _____

3 _____

   _____

4 _____

   _____

5 _____

   _____

6 _____

   _____

7 _____

   _____

8 _____

   _____

9 _____

   _____

10 _____

   _____

11 _____

   _____

12 _____

   _____

13 _____

_____

14 _____

_____

15 _____

_____

2   Action check

SIDE 2

On side two of the audio cassette, you will hear some discussions about safety.

Listen carefully to the extracts and try to answer the questions.

Write your answers and comments in the space below.

Situation 1: _____

_____

_____

_____

_____

_____

Situation 2: _____

_____

_____

_____

_____

_____

Situation 3: _____

_____

_____

_____

_____

Situation 4: _____

_____

_____

_____

_____

_____

3   Unit assessment

Time guide 60 minutes

Read the following case incident and then deal with the questions which follow.

It was freezing in the large temporary building being used as an office by Wilhelmina Spinks and her team of typists. They had moved a week earlier when the weather was warmer. The temperature outside had suddenly dropped, and Wilhelmina realized that there was no heating in the building.

They did have electrical power though, and she went to see her manager about getting some temporary heating fixed up. Between them, they managed to collect together four electrical heaters, two from other departments and two from the maintenance section.

They plugged in the heaters at various points around the office and switched on. There was a bit of a smell of burning dust at first, but the heaters seemed to work all right.

The next morning, Ruth Collins, one of Wilhelmina's team, was first to arrive. It was raining and cold and she was wet through. She switched on all the heaters and sat in front of one of them drying her clothes. She still felt cold, so Ruth decided to move another heater nearer. She picked up a second heater and began to carry it, still switched on, nearer her desk. Because they had only just moved in, the floor was cluttered with files and boxes of paper. Ruth tripped over one of the files and fell. The heater landed on top of her; she was burned and electrocuted. Fortunately, just at that moment two of her colleagues arrived and managed to pull her clear and give her artificial respiration, almost certainly saving her life.

You need only write up to ***three*** or ***four*** sentences in answer to each question.

1. Before the accident occurred, list all the potential hazards that you can spot from the description given.

2. From the description, list all the people who appear to have been at fault in this accident, and the reasons for your answers.

3. Imagine you are investigating the causes of this accident. Write down three key questions you would like to ask any of those involved.

4. From the description, what recommendations would you make to avoid such accidents in future?

4   Work-based assignment

The time guide for this assignment gives you an approximate idea of how long it is likely to take you to write up your findings.

You will find you need to spend some additional time gathering information, perhaps talking to colleagues and thinking about the assignment.

Obtain a copy of your own organization's safety policy statement, and obtain from it answers to the following questions. If you feel that the statement is not clear about any of the points, consult someone who knows the answers – safety officer, safety representative, personnel officer or a suitable member of management.

1. What are the responsibilities of supervisors for health and safety in your organization?

2. What is expected from employees by way of responsibilities and duties for health and safety?

3. What is the declared policy on training? In your opinion, is this policy carried through effectively by you and your colleagues? If it isn't, what more could you do to make the training more effective?

4. What does the policy statement say about communications with employees on health and safety matters? What special arrangements, if any, are there for non-English speaking workers, and for disabled people?

5. How often is the document revised, and how are employees notified about the revisions?

6. How up to date is the document, and to what extent does it take into account recent legislation?

1 Return to objectives

Now that you have completed your work on this unit, let us review each of our unit objectives.

You will be *better able to*:

● recognize the impact of health problems and accidents on working life, and the importance of taking steps to minimize them;

As we have seen, health problems and accidents affect many places of work. No one at work is immune. The published statistics show a frightening picture of injury, illness and fatalities. That's why it's so important for everyone to take a responsible attitude and to play their part in eliminating hazards or in reducing the risk from them.

● explain your duties and responsibilities, and those of your team members, in health and safety matters;

Because a supervisor has to represent the employer, there is a responsibility to carry out the organization's policies and procedures. But the supervisor, like the team member, is also an employee; there are moral and legal obligations for all employees to act in a safe and responsible manner.

● identify the most important legislation related to health and safety and explain the duties imposed by the law on everyone at work;

We have examined the Health and Safety at Work Act 1974 (the HSW Act), the 'six pack' regulations, and the COSHH Regulations. You should make sure you are familiar with the main points of law concerning people at work.

● play an active part in helping the people in your workplace to remain safe and healthy.

I hope you have learned enough from this unit to see ways of improving your own performance, and that of your team, in keeping hazards under control. Remember that health and safety is not something to think about only when something goes wrong, or when the safety officer comes round. Health and safety needs to be at the forefront of your mind during all your working hours.

2 Extensions

Extension 1

Publications of the Health and Safety Executive and the Health and Safety Commission

The address for obtaining these publications, and those in Extension 2, is:

HSE Books
PO Box 1999
Sudbury
Suffolk CO10 6FS
Tel: 0787 881165
Fax: 0787 313995

Alternatively, you can buy them from Dillons Bookstores nationwide or any branch of Rymans the Stationer or Ryman Computer Store.

● *A guide to the Electricity at Work Regulations 1989: an open learning course*, ISBN 0 11 885443 7, £14.00.

● *A guide to the Health and Safety at Work etc. Act 1974: Guidance on the Act*, L1, £4.00.

● *COSHH: an open learning course*, ISBN 0 11 885434 8, £10.00.

● *Display screen equipment (Health and Safety (Display Screen Equipment) Regulations 1992. Guidance on regulations)*, L26, £5.00.

● *Essentials of health and safety at work*, ISBN 0 11 885445 3, £3.50.

● *First aid at work. Health and Safety (First-Aid) Regulations 1981 and guidance*, COP 42, £3.00.

● *Health and safety in wholesale and retail warehouses*, HS(G)76, £7.50.

● *Management of health and safety at work (Management of Health and Safety at Work Regulations 1992. Approved code of practice)*, L21, £5.00.

● *Manual handling (Manual Handling Operations Regulations 1992. Guidance on regulations)*, L23, £5.00.

● *Noise at work*, ISBN 0 7176 0454 3, £3.50.

● *Personal protective equipment at work (Personal Protective Equipment at Work Regulations 1992. Guidance on regulations)*, L25, £5.00.

● The *Safe moves poster No. 6: Safe way: lifting and carrying* is published by the Health and Safety Executive and is available from HSE Books.

● *Safety representatives and safety committees*, ISBN 0 11 883959 4, £2.50.

● *Successful Health and Safety Management*, HS(G)65, £10.00.

● *The costs of accidents at work*, HS(G)96, £8.50.

● *Train to Survive*, C1000, Free.

● *Work equipment (Provision and Use of Work Equipment Regulations 1992. Guidance on regulations)*, L22, £5.00.

● *Work related upper limb disorders: a guide to prevention*, HS(G)60, £3.75.

● *Workplace health, safety and welfare (Workplace (Health, Safety and Welfare) Regulations 1992. Approved Code of Practice)*, ISBN 071 7604136, £5.00.

Extension 2

List of publications on COSHH regulations.

● Health and Safety Executive free leaflets:

   – *Introducing COSHH*;

   – *Introducing Assessment*;

   – *Hazard and Risk Explained*;

   – *COSHH and Section 6 of the Health and Safety at Work Act.*

● Priced booklets:

   – *COSHH Assessments* (a step-by-step guide and the skills needed for it). This is an *aide-mémoire* to employers on what to consider in full assessment where finding out about the risks may not be simple or straightforward.

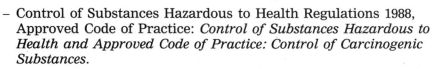

- Control of Substances Hazardous to Health Regulations 1988, Approved Code of Practice: *Control of Substances Hazardous to Health and Approved Code of Practice: Control of Carcinogenic Substances.*

- *A Guide to the Classification, Packaging and Labelling of Dangerous Substances Regulations 1984.*

- Guidance Note EH40/89 (and subsequent editions) *Occupational Exposure Limits.*

- For employers engaged in fumigation operations: *Control of Substances Hazardous to Health Regulations 1988 and Approved Code of Practice: Control of Substances Hazardous to Health in Fumigation Operations.*

- For employers engaged in work with vinyl chloride: *Approved Code of Practice Control of Vinyl Chloride at Work.*

Extension 3

If you have a query about health and safety, the Health and Safety Executive have an information centre. The address is:

HSE Information Centre
Broad Lane
Sheffield S3 7HQ
Tel: 0742 892345
Fax: 0742 892333

Extension 4

Your company will have its own accident report forms. As a supervisor it may well be your responsibility to complete such a form should an accident occur in your section.

You should make yourself familiar with the contents of these forms and it's wise to find out how to fill one out *before* you have to.

An example of an accident report form is shown overleaf:

WAVERY TOWERS
Accident Report Form

1. Employee's particulars:
   Full name (block capitals): ....
   Clock number: ....   Job title...

2. If person not under direct employment, state:
   Full name (block capitals); ....
   Address: ....
   ....

3. Particulars of accident:
   Date: ....   Time: ....   Department: ....
   Time treated: Time sent to *hospital/home/back to work:
   Hospital transferred to: ....*Detained: Yes/No

4. State as accurately as possible nature of injury (state left or right):
   ....

5. State <u>fully</u> treatment rendered: ....
   ....

6. By whom treatment: ....
   ....

7. Particulars of witnesses:
   Full name (block letters): ....
   Address: ....
   ....

8. If no witnesses, who was the first informed? Name:...

9. Description of accident: ....
   ....

10. What specific *job/activity was employee doing at the time of accident? ....
    ....

11. Exact location: ....
    If machine, number and type: ....

12. Was employee authorized or permitted to perform this duty at time of accident? *Yes/No

13. Was employee authorized to be in that place at time of accident? *Yes/No

14. If employee resumed work after accident, state time of finishing shift:
    ....

15. Was accident caused by machinery? *Yes/No
    (a) Was macinery in motion? *Yes/No
    (b) Were all guards in position? *Yes/No
    (c) If YES to (a), state reason why: ....
    If NO to (b), state reason why: ....

16. Date ceased work: ....   Time: ....

17. Was anyone else involved? If so, state name: ....
    Clock No: ....
    By whom employed: ....

    Department: ....
    Signatures:
    Medical department: ....   Date: ....
    Supervisor: ....   Date: ....
    Manager: ....   Date: ....
    Safety Officer: ....   Date: ....

Please return this form to the safety officer as soon as it is completed.

*Delete as appropriate.

These Extensions can be taken up via your Support Centre. They will arrange for you to have access to them. However, it may be more convenient to check out the materials with your personnel or training people at work – they could well give you access. There are good reasons for approaching your own people as, by doing so, they will become aware of your continuing interest in the subject and you will be able to involve them in your development.

ACTION PLAN

Action

Work out your own plan of action for improving the health and safety record of your work area and the awareness of your team regarding matters of health and safety by responding to the following check questions and picking up the action points.

Check question

Your response and intended action:

1 How well do you know the law on health and safety?

_____

_____

■ *If you are still not sure about some of the points of law affecting you and your job, make it your business to find out. Check with the people in your organization with special responsibility for health and safety.*

2 To what extent are you sure that you are conforming with the law in all respects?

_____

_____

■ *The booklets published by HSE on the new 'six pack' regulations give guidance, and explain the codes of practice. Ask your employer for copies, unless you are given this information in another way. Remember – you could be personally liable for prosecution if you break the law!*

3 How well do you know your own organization's policy, system and detailed arrangements for health and safety?

_____

_____

■ *Make sure you understand them, particularly in so far as they affect you and your team. Ask for clarification if you're not sure.*

4 How well-informed are your team members about health and safety matters?

_____

_____

■ *It's part of your job to keep the team informed about matters of health, safety and welfare. You could start by trying to write down what they need to know. You may need a lot more space than we give here!*

5 How often do you carry out regular and systematic checks and inspections of safety arrangements, equipment etc?

_____

_____

■ *Unless you have already done so, why not draw up a schedule now? Talk to your boss about it.*

6 To what extent do you insist on every team member using the correct personal protective equipment at all times?

_____

_____

■ *Your insistence on team members sticking to the rules could save lives.*

85

7 How much training do you provide in health and safety topics?

_____

_____

■ *It is part of a supervisor's job to see that the workteam are fully trained to be safe.*

8 What do you do when you see someone breaking the safety rules – whether or not they are in your group?

_____

_____

■ *You know what you should do – take action! When someone breaks the rules others tend to follow. Sooner or later somebody will get hurt.*

9 Do you consistently set a good example on health and safety?

_____

_____

■ *Your team need to see you following the rules, as well as reciting them. Don't forget also that supervisors are often among those most at risk.*

10 What hazards exist in your work area?

_____

_____

■ *If you haven't already done so, take time to identify the hazards. Get help on this.*

11 To what extent have the risks from these hazards been assessed?

_____

_____

■ *Risk assessment is required by law. Work with your colleagues on this.*

12 What will happen in the event of an emergency, such as a fire? Do your team know exactly what to do?

_____

_____

■ *Tighten up on your procedures. Make sure everyone knows what to do, where the emergency equipment is, etc. Get help if you aren't sure about this.*

13 Are the members of your team quite clear about the procedures for reporting accidents?

_____

_____

■ *It might be an idea to run through the procedures with them.*

14 To what extent do you contribute to creating and maintaining a positive culture for health and safety?

_____

_____

■ *Remember the four Cs: competence, control, cooperation and communication.*